ENDOCRINES AND ENZYMES IN ANESTHESIOLOGY

The Proceedings of the Eleventh Biennial Western Conference on Anesthesiology

ENDOCRINES AND ENZYMES IN ANESTHESIOLOGY

Edited by

CARTER M. BALLINGER, M.D.
Chairman
Division of Anesthesiology
University of Utah Medical Center
Salt Lake City, Utah

and

VERNE L. BRECHNER, M.D.
Professor and Vice Chairman
Department of Anesthesiology
UCLA School of Medicine
Los Angeles, California

CHARLES C THOMAS • PUBLISHER
Springfield • Illinois • U.S.A.

Published and Distributed Throughout the World by
CHARLES C THOMAS • PUBLISHER
BANNERSTONE HOUSE
301–327 East Lawrence Avenue, Springfield, Illinois, U.S.A.

ISBN 0–398–02463–4

Library of Congress Catalog Card Number: 73–187643

With THOMAS BOOKS *careful attention is given to all details of manufacturing and design. It is the Publisher's desire to present books that are satisfactory as to their physical qualities and artistic possibilities and appropriate for their particular use.* THOMAS BOOKS *will be true to those laws of quality that assure a good name and good will.*

Printed in the United States of America

BB-14

Contributors

EUGENE ACKERMAN, Ph.D.

Hill Family Professor and Director
Division of Health Computer Sciences
Department of Laboratory Medicine
University of Minnesota
Minneapolis, Minnesota

GUILLERMO M. CREMER, M.D.

Resident, Department of Internal Medicine
Mayo Graduate School of Medicine
University of Minnesota
Rochester, Minnesota

STUART C. CULLEN, M.D.

Professor of Anesthesiology
School of Medicine
University of California
San Francisco, California

BROWN M. DOBYNS, M.D.

Professor of Surgery
School of Medicine
Case Western Reserve University
Cleveland Metropolitan General Hospital
Cleveland, Ohio

A.H. GIESECKE, JR., M.D.

Professor of Anesthesiology
Southwestern Medical School
University of Texas
Parkland Memorial Hospital
Dallas, Texas

AARON J. GISSEN, M.D.

Department of Anesthesiology
Columbia University
New York, New York

DAVID YI-YUNG HSIA, M.D.

Professor and Head, Department of Pediatrics
Stritch School of Medicine
Loyola University
Hines, Illinois

M.T. JENKINS, M.D.

Professor and Head, Department of Anesthesia
Southwestern Medical School
University of Texas
Parkland Memorial Hospital
Dallas, Texas

WERNER KALOW, M.D.

Professor and Chairman, Department of Pharmacology
Faculty of Medicine
University of Toronto
Toronto, Canada

JOANNES H. KARIS, M.D.

Department of Anesthesiology
Columbia University
New York, New York

COLONEL JAMES S. KETCHUM

Chief, Behavioral Science Division
Medical Field Service School
Fort Sam Houston, Texas

GEORGE D. MOLNAR, M.D., Ph.D.

Professor of Internal Medicine
Mayo Graduate School of Medicine
University of Minnesota
Rochester, Minnesota

KAREN E. MOXNESS, M.S.

Nutritionist
Mayo Clinic
Rochester, Minnesota

WILLIAM L. NASTUK, Ph.D.

Professor of Physiology
College of Physicians and Surgeons
Columbia University
New York, New York

DON H. NELSON, M.D.

Professor of Medicine
University of Utah
Chief of Medicine
Latter-Day Saints Hospital
Salt Lake City, Utah

TSUTOMU OYAMA, M.D.

Professor and Chairman, Department of Anesthesia
Hirosaki University
School of Medicine
Hirosaki, Aomori-Ken, Japan

THEODORE ROBERTS, M.D.

Associate Professor of Surgery
Chairman, Division of Neurosurgery
School of Medicine
University of Utah
Salt Lake City, Utah

JOHN W. ROSEVEAR, M.D., Ph.D.

Clinical Associate Professor
of Laboratory Medicine
Department of Biochemistry
University of Minnesota
Minneapolis, Minnesota

OLGA SCHWEIZER, M.D.

Chief of Service
Chief, Department of Anesthesiology
Memorial Hospital for Cancer and Allied Diseases
New York, New York

F. JOHN SERVICE, M.D., Ph.D.

Instructor in Internal Medicine
Mayo Graduate School of Medicine
University of Minnesota
Rocthester, Minnesota

Preface

In 1949, the California Society of Anesthesiologists organized a conference on anesthesiology and presented the subject of respiration. The conference rapidly expanded to include other Western states and is now known as the Biennial Western Conference on Anesthesiology.

Since the initial meeting, many subjects have been discussed; for example, circulation and the nervous system. This year the program committee chose a topic which had been a previous theme, endocrines and enzymes. Because of the significant changes that have been made during the fifteen-year interim, the topic is quite appropriate.

The subject was thoroughly covered by the competent participants—from Doctor Kalow's opening discussion of enzyme induction by barbiturates, to Doctor Jenkins' closing description of exaggerated hormonal responses during anesthesia.

It is the hope of those who took part in this eleventh conference that the published proceedings will be interesting and beneficial to all academic *and* practicing anesthesiologists.

CONTENTS

ENDOCRINES AND ENZYMES IN ANESTHESIOLOGY

Chapter 1

Enzyme Induction by Barbiturates and Other Drugs

WERNER KALOW

During the past decade, it has become apparent that drugs may stimulate the biotransformation of drugs in animals and in man.[1-6] I wish to describe some of the pertinent observations, first by presenting a number of animal data, second by reviewing information concerning man, and third, by giving a brief account of studies designed to elucidate mechanism and nature of the effects under discussion. I will not try to cover the molecular aspects in any detail. I hope you will be able to assess afterwards for yourself whether this phenomenon deserves your attention.

Everyone who gives drugs is aware of a person-to-person variation in the response. Some people get more easily drunk than others, and some become addicted to alcohol while others do not. Some persons need more thiopental than others. Some may be helped by one antihistaminic more than by another. Some patients tend to get a toxic side effect from a drug which is well tolerated by others. It appears that variation in drug metabolism accounts for much of the variation in drug response. The rate of drug metabolism is often assessed by observing levels of drugs in blood or plasma. The magnitude of the problem is illustrated by the widespread experience that blood levels of drugs after standard intake can be expected to vary among patients by a factor of 3 even in the absence of known complicating factors.

ADAPTIVE PHENOMENA IN ANIMALS

One reason for differences between individuals may be a different state of adaption to a drug. One of the early examples of adaptation was observed by Remmer[5] on hexobarbital, also known as Evipal®, an ultra short-acting barbiturate widely used in Europe. A dog was given a standard injection of hexobarbital on five consecutive days. The drug levels were measured over two and one-half hours following each injection, including the time when the animal woke up. On the first day, the animal slept for almost two hours and woke up when the blood level was between 25μg and 30μg/ml of blood. On the second day, the animal slept for only an hour, although the blood level at the time of awakening was again 25μg/ml. On the third and the fifth days, blood levels at awakening were again similar, but the rate of hexobarbital elimination was even faster and the sleeping time was only about forty-five minutes. Four weeks later the dog was almost back to the original response. If one had simply measured sleeping time, one would say the dog developed tolerance to hexobarbital by repeated injection and this tolerance was produced by an increased rate of the elimination of the drug.

Another aspect of the same principle is given with data on tolbutamide[5], an oral antidiabetic. Dogs received a tablet once a day for a few weeks. Initially the drug levels in serum were about 30 mg/100 ml; a week later they were down to about 10 mg/100 ml. The increased rat of elimination led to a decreased cumulation of the drug in the body. This is again an example of the development of tolerance because of an adaptation which consisted in an increased metabolism of the drug.

A different aspect of the same principle is revealed by the stimulation of the metabolism of one drug by the administration of another. A set of classical data showed blood levels of diphenylhydantoin (Dilantin®) after intravenous injection in two dogs. Both dogs were then given a barbiturate consecutively for a few days, after which time the intravenous injection of diphenylhydantoin was repeated. In both dogs the barbiturate pretreatment caused a drastic acceleration of the disappearance of diphenylhydantoin from the plasma. The half-life of diphenylhydantoin in plasma was decreased by almost an order of magnitude.[8]

Some synthetic chemicals which are not drugs or medicinal agents, are nevertheless able to stimulate drug metabolism.[7,12] For instance, the insecticide chlordane, which is contained in numerous commercial preparations mixed with other ingredients, and which is used in households against cockroaches, is one of the most effective agents in shortening the action of drugs. Most remarkable is the persistence of the effect. When rats received phenylbutazone, an anti-inflammatory agent first marked as Butazolidin®, the drug levels in their blood were still substantially reduced sixteen weeks after termination of the application of chlordane.

Phenobarbital is capable of accelerating the metabolism not only of drugs but also of some hormones.[3] In the rat, this has been shown for progesterone, deoxycorticosterone, androsterone, and various estrogens. The implications of these observations in medicine are under vigorous investigation.

Vesell[19] and his colleagues showed that the bedding in the cages of mice influenced the effect of hexobarbital. When wood shavings of cedar were used as the bedding, the sleeping time shortened drastically, while the ***in vitro*** ability of the livers of these mice to metabolize hexobarbital increased. The changeover to shavings of hardwoods reversed this process. The agent responsible for this effect is not known, but this kind of experience could cause a pharmacologist to turn grey.

The authors also tested enzyme activity in what they call "9000 gm supernatant" of liver homogenates. This is another way of saying, they tested the capacity of the liver *microsomes* of the mice to metabolize the drug.

I wish to reinforce this emphasis on microsomes by giving another example. The study involved the *in vitro* metabolism of the anticoagulant bishydroxycoumarin (dicoumarol)[8]. When young male rats were pretreated for four days by intraperitoneal injections of phenobarbital, the isolated liver microsomes of these rats were almost thirty times more effective in metabolizing dicoumarol than were the microsomes of untreated rats. Thus, when considering drug metabolism and the disappearance of drugs from the plasma, one must pay special attention to the enzymes located in that particular cell fraction of the liver which is commonly designated as endoplasmic reticulum or—after cell fractionation and centrifugation—as microsomal fraction.

Let this account of experimental protocols in animals suffice. We have seen that drugs and chemicals can cause a change in liver microsomes which leads to an increased rate of metabolism of some drugs. I shall leave it to later to explain why this is referred to as adaptive enzyme formation or enzyme induction rather than simply enzyme activation.

Although a drug may stimulate its own metabolism, the ability of drugs to induce the formation of microsomal enzymes nevertheless is not related to the metabolism of these drugs. For instance, phenobarbital is a good stimulant of enzyme formation although it is metabolized slowly. Indeed it has been repeatedly observed that there are specific structural requirements for this stimulating effect on enzyme formation. Table 1-I shows, for instance, how of pairs of chemically similar drugs, one is an enzyme inducer, the other not. There are many dozens of drugs which

TABLE 1-I
CONTRASTING POTENCIES OF ENZYME INDUCTION OF RELATED AGENTS

Glutethimide (Doriden)	+++	Thalidomide	0
Paramethadione (Paradione ®)	+++	Trimethadione (Tridione ®)	0
Chlorpromazine	++	Promazine	0
Tolbutamide (Rastinon ®, Orinase ®)	+++	Sulfanilamide	0
Chloroform	+	Halothane	0

have been shown to induce microsomal enzymes although not every inducing drug does so in all species.

Some of the most potent inducers are shown in Table 1-II. Phenobarbital is the best known and the most widely used inducing agent. The most potent inducers of all are the insecticides of the chlorinated hydrocarbon type, including DDT. Some cancerogenic materials like methylcholanthrene or benzyprine must differ somewhat in the mechanism of action from

TABLE 1–II
THE MOST POTENT INDUCERS

Phenobarbital Phenylbutazone Nikethamide		Glutethimide Phenaglycodol Tolbutamide
Chlordane	Lindane	DDT
Methylcholanthrene		Benzpyrene

the others in that they induce a slightly different group of enzymes. I can only mention in passing that there are a number of drugs which have often been tested, and yet do not enhance enzyme activity. Among these are, for instance, morphine, Demerol®, atropine, and similar agents.

Microsomal enzymes which metabolize foreign compounds are numerous or have numerous capabilities. Table 1-III summarizes reactions catalized by microsomes and some representative drugs affected by them.

TABLE 1–III
REACTIONS CATALYZED BY MICROSOMAL ENZYMES WHICH METABOLIZE FOREIGN COMPOUNDS

1. Oxidations	
Aromatic hydroxylation	diphenylhydantoin, zoxazolamine, ethylbiscoumacetate, acetanilide, steroids
Aliphatic hydroxylation	hexobarbital, tolbutamide
N-Dealkylation	nicotine, tremorin, meperidine, amidopyrine
O-Dealkylation	codeine, phenacetin
Deamination	amphetamine
Sulfoxidation	chlorpromazine
N-Oxidation	aniline
S-Exchange	parathion
2. Reductions	
Conversion of nitro- to aminogroup by nitroreductase	chloramphenicol, nitrobenzene
Splitting of azo group by azoreductase	prontosil
3. Hydrolysis	
Splitting of ester	procaine
4. Conjugations	
Glucuronic acid conjugation	sulfadimethoxine
Mercapturic acid synthesis (acetyl cysteine conjugation)	naphthalene

The largest number of reactions are oxidation reactions. Most of the previous examples had to do with hydroxylation reactions. I should perhaps mention that procaine is hydrolyzed in the rat by a microsomal enzyme, but in man by the cholinesterase in plasma. In this respect man is different from most or all other species.

EXAMPLES OF ADAPTIVE ENZYME FORMATION IN MAN

Cucinell *et al.*[8] tested the plasma levels of dicumarol in five subjects after a standard dose. Dicoumarol was given again and the measurements repeated after a series of treatments with phenobarbital. In four of the five persons, the plasma level of dicoumarol was much lower after phenobarbital treatment than before. The one exception was a subject who had eliminated dicourmarol at a fast rate prior to the administration of phenobarbital. This experiment also has been performed under clinical conditions. Data obtained on a single person who was for many months on dicumarol treatment[8] showed that there were two periods during which he received phenobarbital in addition to the dicourmarol. Both the prothrombin time and the dicourmarol blood level decreased sharply during these periods.

This interaction has caused trouble on more than one occasion. A person entering hospital after a cardiac attack is often given a barbiturate as sedative, and an anticoagulant. Prothrombin time is checked daily and the dose of the anticoagulant adjusted while the patient is in hospital. The patient is then discharged, and the barbiturate intake stops; the prothrombin time immediately gets off control because of the rise of anticoagulant level. The patient may die of bleeding. Vice versa, a patient adjusted to the anticoagulant warfarin without barbiturate was given a barbiturate as a preoperative sedative. An enhanced elimination of warfarin most likely contributed to the ensuing fatal thrombosis.

It is almost routine therapy to give an epileptic patient phenobarbital as well as Dilantin. What one does by using this combination is not yet entirely clear. It has been shown that the Dilantin level in plasma decreases substantially if phenobarbital is added to the regimen.[8] A recent report printed only in the form of an abstract suggests that this does not occur in all patients.[9] Since the therapeutic effects of the two drugs are similar, an epileptic attack may be suppressed regardless of some alteration of the metabolism of diphenylhydantoin. However, it follows that for optimal therapy and to avoid gum hyperplasia, one may have to determine blood levels of these drugs in a patient in order to adjust dosage.

Hammer and Sjøqvist[14] have shown by an experimental procedure in one subject that levels of the antidepressant desipramine in plasma were lower after phenobarbital than before. However, even the phenobarbital-stimulated metabolism did not seem to be nearly as fast as that occurring spontaneously in some subjects. Remmer[5] has tested drug metabolism in subjects one week and three weeks after attempted suicides with an overdose with a hypnotic. This indicated that various barbiturates as well as glutethimide (Doriden®) could enhance microsomal reactions.

A recent investigation dealt with the stimulating but highly variable effect of smoking on benzpyrene hydroxylase. This could have some bearing on susceptibility to cancer, since the benzyprenes are carcinogens.[3]

Not very much is known about the influence of disease on drug metabolism. A study by Levi *et al.* suggests that liver disease may reduce the activity of microsomal enzymes on the average.[13] However, in spite of liver disease, barbiturates still enhanced enzyme activity. More work should be done in this area.

THE MOLECULAR PHARMACOLOGY OF ENZYME INDUCTION

The Nobel Prize-winning concepts of Jacob and Monod on induction of bacterial enzymes should not be directly transferred to drug-stimulated adaptive enzyme formation in man or in mammals.[10]

Rats subjected to chronic treatment with phenobarbital and other enzyme inducers increase their liver weight relative to body weight.[5] Part of this increase seems to be due to a change in cytoplasm, in nuclei and in mitochondria, but the only big increase is in the so-called smooth microsomes. In this microsomal fraction one can show a doubling of protein content and a modest increase in phospholipids and ribonucleic acid. The stimulating action of phenobarbital obviously is not a simple process. However, there is a more than threefold increase in one particular protein, namely a cytochrome which is referred to as cytochrome P-450.[5,6] As all cytochromes, it is a heme compound and can therefore be spectrophotometrically measured with much greater precision and specificity than other proteins. One experimental difficulty is, however, that the microsomes contain another cytochrome referred to as cytochrome B_5, which is far less affected by drugs than is cytochrome P-450.

Two years ago, Schenkman, Remmer, and Estabrook[16] made an observation which is still puzzling many pharmacologists and biochemists. On addition of drugs to microsomes, the cytochrome P-450 changes its absorption spectrum, indicating that the cytochrome and the drug form a complex. Indeed different drugs form two different kinds of complexes which have been designated as Type 1 and Type 2; prototypes are the complexes with hexobarbital and aniline, respectively. What this means remains to be investigated. Are there several cytochromes P-450?[11] Is the cytochrome-drug complex combining with other enzymes? Any postulate will have to account for some peculiar sex-dependent differences in drug metabolism and also for the selectivity in adaptive phenomena.[15] Several groups of investigators are trying to solve these and related problems.

OUTLOOK AND CONCLUSIONS

I have tried to convey a broad picture of what is presently known about enzyme induction by drugs. I have not commented on enzyme induction by hormones.[18] I wish to conclude by putting the phenomenon of enzyme adaptation to drugs into pharmacological perspective.

If one drug *enhances* the metabolism of another, this is only *one* particular aspect of drug interaction. Even in the field of drug metabolism, it often occurs that one drug *inhibits* the metabolism of another, rather than accelerating it. In addition, drugs may interact at the target organ, to produce synergism or antagonism unrelated to any question of drug metabolism. In recent years plasma binding of drugs has been found to be an important factor in drug distribution in the body; that is, plasma albumin behaves like a major reservoir capable of carrying substantial amounts of drug. Drugs may compete for space in this reservoir, and one may drive out the other.[17] Thus, adaptive enzyme formation is only one facet of drug interactions which are intensely studied by numerous investigators.

Another problem relates to the question of how much of the individual variation in drug metabolism has a hereditary cause and how much is due to adaptive changes produced by drugs or other chemicals. A recent series of investigation on twins, by Vesell and his collaborators at the National Institutes of Health, suggest that heredity plays the primary role in drug elimination and accounts for most of what may be called normal variation. However, we are not able to fully assess the relative importance of heredity and adaptation. Both factors are operative, and both may be important simultaneously in that the adaptive enzyme formation can take place only in the presence of the right kind of genes.

Finally, one should mention the investigations designed to find out whether the stimulating activity of phenobarbital on the liver can be utilized therapeutically to overcome defects of glucuronidation, or to prevent or ameliorate liver damage due to disease or due to toxic agents.

In any case, it is encouraging to see that this new aspect of pharmacology which has been developed within a very few years leads one to envision new therapeutic possibilities.

BIBLIOGRAPHY

Reviews and General References

1. Burns, J.J., and Conney, A.H.: Enzyme stimulation and inhibition in the metabolism of drugs. *Proc Roy Soc Med, 58*:955, 1965.
2. Conney, A.H.: Pharmacological implications of microsomal enzyme induction. *Pharmacol Rev, 19*:317, 1967.
3. Conney, A.H.: Drug metabolism and therapeutics. *N Engl J Med, 280*:653, 1969.
4. Greene, N.M.: The metabolism of drugs employed in anesthesia. *Anesthesiology, 29*:127, 1968.
5. Remmer, H.: Drug-induced formation of smooth endoplasmic reticulum and drug metabolizing enzymes. *Proc Europ Soc Study Drug Toxic, 4*:57, 1964.
6. Remmer, H.; Estabrook, R.W.; Schenkman, J., and Greim, H.: Reaction of drugs with microsomal liver hydroxylase: its influence on drug action. *Arch Pharmak Exp Path, 259*:98, 1968.

Papers Cited

7. Conney, A.H.: Effects of pesticides on drug and steroid metabolism. *Clin Pharmacol Ther, 8*:2, 1967.
8. Cucinell, S.A.; Conney, A.H.; Sansur, M., and Burns, J.J.: Drug interactions in man. *Clin Pharmacol Ther, 6*:420, 1965.
9. Frey, H.H.; Kampmann, E., and Nielsen, C.K.: Experimental study on combined treatment with phenobarbitone and diphenylhydantoin. *Acta Pharmacol Toxic 25* (Suppl. 4):66, 1967.
10. Gelboin, H.V.: Drugs and Protein Synthesis. *23 Suppl.*, 85, 1965.
11. Hildebrandt, A.; Remmer, H. and Estabrook, R.W.: *Biochem Biophys Res Commun, 30*:607, 1968.
12. Juchau, M.R.; Gram, T.E., and Fouts, J.R.: Stimulation of hepatic microsomal drug-metabolizing enzyme systems in primates by DDT. *Gastroenterology, 51*:213, 1966.
13. Levi, A.J.; Sherlock, S., and Walker, D.: Phenylbutazone and isoniazid metabolism in patients with liver disease in relation to previous drug therapy. *Lancet, I*:1275, 1968.
14. Hammer, W., and Sjoqvist, F.: Estimation of Monomethylated Tricyclic Antidepressants by Acetylation with ^{3}H-Acetic Anhydride—Methodology and Application in Clinical Pharmacology. Workshop on Biochemical Approaches to Clinical Pharmacology, Vanderbilt University, June, 1967, Nashville, Tennessee.
15. Schenkman, J.B.; Frey, I.; Remmer, H., and Estabrook, R.W.: Sex differences in drug metabolism by rat liver microsomes. *Mol Pharmacol, 3*:516, 1967.
16. Schenkman, J.B.; Remmer, H., and Estabrook, R.W.: Spectral studies of drug interaction with hepatic microsomal cytochrome. *Mol Pharmacol, 3*:113, 1967.
17. Solomon, H.M.; Schrogie, J.J., and Williams, D.: The displacement of phenylbutazone-^{14}C and warfarin-^{14}C from human albumin by various drugs and fatty acids. *Biochem Pharmacol, 17*:143, 1968.
18. Weber, G.; Singhal, R.L.; Stamm, N.B., and Srivastava, S.K.: Hormonal induction and suppression of liver enzyme biosynthesis. *Fed Proc, 24*:745, 1965.
19. Vesell, E.S.: Induction of drug-metabolizing enzymes in liver microsomes of mice and rats by softwood bedding. *Science, 157*:1057, 1967.

Chapter 2

Human Developmental Genetics

DAVID YI-YUNG HSIA

During the past decade, we have witnessed tremendous progress in the areas of human biochemical genetics and chromosomal aberrations in men. These studies have dealt with genetic defects which become manifest after the birth of an infant and which sometimes do not become apparent until much later in adult life. Fortunately, it has become increasingly clear that genetic influences play an important role in the growth and development of the fetus during intrauterine life, and that we should think more in terms of factors that cause perinatal morbidity and mortality. Human developmental genetics may be considered under five headings: (1) genetics of prenatal growth (2) cytogenetics, (3) biochemical genetics, (4) immunogenetics, and (5) iatrogenic factors.

GENETICS OF PRENATAL GROWTH

In this section we shall examine the genetic and environmental factors that play roles in determining the physical growth and development of the human fetus. It is apparent that a physical measurement such as birth weight does not follow the mendelian concept of single gene inheritance. We know that the mother does not possess a single "gene" which controls the size of the infant she will have, just as the father does not have a "gene" that governs the height of his daughters. However, there is good reason to believe that a group of genes together play a role in determining certain physical measurements and that these interact with external variables to produce the final result. We believe that this concept of polygenic or multifactoral inheritance has been somewhat neglected by obstetricians and pediatricians and the present review represents a summary of what is known at present about the genetics of prenatal growth.

The concept of multifactoral inheritance was originally formulated by Mendel himself. When he crossed two varieties of flowering beans, one having purple flowers and the other white flowers, he found that the F_2 generation had a range of colors from purple to white. He suggested that

These studies were aided by grants from the Illinois Mental Health Fund and the United States Public Health Service (HD 04346).

this might be due to additive action of more than one gene. In 1918, Fisher very simply showed that the regression of child on parent, or parent on child, or sibling on sibling is equal to the number of genes in common and this is also true for more remote relationships as illustrated in Table 2-I. We can illustrate this by suggesting that body height is determined by a combined action of ten separate and independent genes. In an identical twin, one would expect all ten genes to be the same and these

TABLE 2–I
GENES COMMON IN MULTIFACTORIAL INHERITANCE

Relationship to a Given Subject	*Proportion of Genes in Common*
Identical twin	1
Fraternal twin	1/2
Parent, child, sib	1/2
Grandparent, grandchild, uncle, aunt, nephew, niece, half-sib	1/4
First cousin	1/8
Second cousin	1/32

From *Human Developmental Genetics*, by David Yi-Yung Hsia. Copyright © 1968 Year Book Medical Publishers, Inc. Used by permission.

would cause the second twin to have essentially the same height as the first twin. In the case of fraternal twins, brothers, sisters, parents, and children, one would, on the average, expect to have five genes which are the same and five other genes which are no more alike than that of an unrelated person. In these individuals, instead of half of the siblings, parents, children being totally alike, and half totally different, they tend to be half alike. Thus, the closer the degree of relationship, the more similar the trait. One of the best examples of multifactoral inheritance is found in the studies by Holt of fingerprint ridge patterns, which are shown in Table 2-II. One great advantage of this characteristic is that

TABLE 2–II
FINGERPRINT RIDGE COUNTS IN RESEMBLANCES BETWEEN RELATIVES

Relationship	*No. of Pairs*	*Observed Correlation*	*Theoretical Correlation*
Identical twins	80	0.95 ± 0.01	1.0
Fraternal twins	92	0.49 ± 0.08	0.5
Sib-sib	642	0.49 ± 0.04	0.5
Parent-child	602	0.50 ± 0.03	0.5
Husband-wife	149	0.05 ± 0.08	0.0

From *Human Developmental Genetics*, by David Yi-Yung Hsia. Copyright © 1968 Year Book Medical Publishers, Inc. Used by permission.

the number of ridges is finally determined by the fourth month of fetal life and is little affected by the length of gestation and environmental changes thereafter. It can be seen that there is almost a perfect fit between the observed correlation and theoretical correlation in various degrees of relative study.

It is obvious that the intrauterine growth of the fetus is also influenced greatly by environmental factors. For example, nutrition, oxygenation, smoking, infections, and placental function have been shown to influence the size of the baby. We are just beginning to understand the complex interplay between genetics and environment, and methods are beginning to be developed to assess the importance of each separate factor on normal and abnormal growth.

In Table 2-III, Penrose has set up the first approximation for the hereditary and environmental factors that determine birth weight. They are based on a series of arithmetic calculations made from data collected from British newborns. He starts off by pointing out that a significant likeness was found among children of sisters, but not among children

TABLE 2-III
APPROXIMATE PARTITION OF CAUSES OF VARIATION OF BIRTH WEIGHT BETWEEN HEREDITARY AND ENVIRONMENTAL CAUSES

	Per Cent
Maternal hereditary constitution	20
Fetal hereditary factors: sex	2
Fetal hereditary factors: remaining constitution	16
Maternal general health and nutrition	16
Maternal health during particular pregnancy	8
Maternal parity	7
Maternal age	1
Unidentified intrauterine influence, posture, etc.	30

From *Human Developmental Genetics*, by David Yi-Yung Hsia. Copyright © 1968 Year Book Medical Publishers, Inc. Used by permission.

of brothers, or brother and sister groups, indicating that the hereditary constitution of the mother is more important than the father. The correlation of birth weight between cousin pairs where mothers are sisters is $0.13 + 0.04$. If one estimates that the quantity of maternal hereditary effect is no greater than twice the correlation of the sisters' children, it is estimated at about 20 per cent after allowances have been made for similarity of the fetal constitution of cousins. Penrose then compares the correlation for monozygotic and dizygotic twins and finds that this may differ by 0.09 ($0.67 - 0.58 = 0.09$); when doubled, this gives an estimate of the effect of fetal constitution of 18 per cent. This particular estimate includes the variations due to sex, which amount to some 2 per cent, and thus, hereditary factors excluding sex amount to 16 per cent.

The effect of birth order and maternal age can be independently measured, and they amount to 7 per cent, and 1 per cent respectively. The nutritional status and the general health of the mother (not specific for any pregnancy) can be estimated to be 16 per cent, and maternal health during each particular pregnancy can be found to be 8 per cent. The remaining 30 per cent must be attributed to identified intrauterine influences, posture, etc. Thus, one may view the 20 per cent due to maternal hereditary constitution, the 16 per cent due to fetal hereditary constitution, and the 2 per cent due to sex, for a total of 38 per cent of the variations in birth due to heredity. Similarly, we can consider the 16 per cent due to maternal general health, the 8 per cent due to maternal health during pregnancy, the 7 per cent due to parity, the 1 per cent due to maternal age, and the 30 per cent due to unidentified intrauterine influence, or a total of 62 per cent of the variations in birth weight due to environment. This is an extremely crude first approximation, but it does demonstrate that scientific and mathematical methods can be usefully brought to play in estimating the variation of intrauterine growth.

CYTOGENETICS

Accurate knowledge of the number of chromosomes and their behavior in mitosis and meiosis has contributed immeasurably to our understanding of the development of the human fetus. Until recently, the scope of human cytogenics was limited to attempting to count accurately the number of human chromosomes and to verifying that, in the human species as in other mammalian species, there was a sexual dichotomy in the chromosomes. Normal males were found to have an X and a Y chromosome, and normal females were found to have two X chromosomes. In 1949, Murray Barr and one of his graduate students, Ewart Bertram, observed a difference in the morphology of cell nuclei, depending on the sex of the individual providing the tissue. In normal females, it was found that there is a distinct chromatic body which usually adheres closely to the inner surface of the nuclear membrane. In normal males, this body is absent. More recent work suggests that the sex chromatic body represents the heterochromatic portion of one or both X chromosomes. In 1956, Tijo and Levan, using the technique of growing human cells in culture in the presence of colchicine (which inhibits the formation of the mitotic spindle, arrested mitosis, and accumulated cells at metaphase), were able to show that the chromosome count in man was 46. This was composed of 44 autosomes and 2 sex chromosomes and was contrary to the count of 48 chromosomes which had been accepted by all workers since 1923, when Painter reported that number from his studies in normal testicular tissues. In 1959, Lejeune, Gautier, and Turpin described a consistently abnormal chromosome number in the somatic cells of "mongoloid" idiots. By preparing the karyotype, it can be seen that these individuals have three rather than two of the small acrocentric chromosomes with satellite bodies. This extra chromosome has been designated as number 21 by common agreement. These three milestones have led to an understanding of a formerly unknown mechanism in the causation of human disease.

In 1949, Hertig and Rock found pathological abnormalities in twelve of twenty-eight early human conceptuses in uteri that were therapeutically removed for a variety of reasons. Five of these showed minor degrees of trophoblastic hypoplasia, malorientation of the embryo, and various degrees of shallow implantation. These five might or might not have aborted, probably not. The remaining seven showed defective blastomeres, trophoblast only, absent chorionic cavity, and markedly defective trophoblast. All of these would have aborted, but in four instances the abortion may have been missed, since there may have been no clinical evidence of pregnancy in the mother because of their extreme abnormality. These classic studies led to much speculation as to whether miscarriages were caused by a defective implantation in the mother or whether they represent nature's way of getting rid of a defective ovum. This controversy was resolved in a most dramatic fashion by Penrose and Delhanty finding a triploidy with an XXY constitution in material obtained from spontaneous abortion.

It should be stressed that only about half of the aborted material will grow in tissue culture and yield cells which can be successfully examined for chromosomes. This is caused partly by the fact that many of the conceptuses have been dead for some time prior to the miscarriage and partly because there usually is some delay between collecting the material and starting the culture. Despite these difficulties, a surprising amount of information has been collected during the past few years.

Studies using sex chromatin tests have shown that the ratio of males to females is about 106 : 100 at the time of birth and probably closer to 130 : 100 during the early part of pregnancy. As shown in the left hand side of Table 2-IV, chromosome studies in material obtained by therapeutic abortion showed 79 males and 71 females, or a sex ratio of 111 : 100. This is consistent with the findings determined from sex chromatin. However, chromosome studies on material obtained from spontaneous abortions showed 112 males and 133 females, or a ratio of 83 : 100. This excess of females has led to the speculation that in some instances one is examining placental tissue rather than embryonic tissue. Recently, it has been suggested that fetal membranes should be grown in culture by placing three to four explants between two glass slips without the use of clot and trypsin. This differentiates between placental tissue, which has a much shorter life span, and embryonic tissue, which has a much longer life span.

There is a striking difference in the frequency of chromosomal abnormalities in embryos obtained from therapeutic abortions and from

TABLE 2–IV

CHROMOSOMAL ABNORMALITIES IN HUMAN ABORTIONS

Type of Abortion: *Therapeutic* Males	*Therapeutic* Females	*Therapeutic* Total No.	*Therapeutic* Abnormal No.	*Spontaneous* Males	*Spontaneous* Females	*Spontaneous* Total No.	*Spontaneous* Abnormal No.	Chromosome Abnormality: *Trisomy* A	B	C	D	E	F	G	*Sex Chromosome* XO	XXY	*Triploidy* XXY	XYY	XXX	*Tetraploidy*	XXXX	XYYY	*Others*	*Total*
63	64	127	2																				2[1]	2
				5	4^+	10	3	1[2]				1			1									3
				4	4	8	2	2																2
5	3	8	2[3]	5	6^+	12	7					3		1	1				1		1	1[4]	1[3]	9
11	4	15	0	10	9^+	25	16			1		2		3	5		4			1				16
				85	104^+	200	44		2	2	6	7		5	11		6	1	2		1	1		44
				3	6	9	1										1[5]							1
79	71	150	4 (2.6%)	112	133	264	73 (27.6%)	3	2	3	6	13	0	9	18	0	11	1	3	1	2	2	3	77

$^+$ Excludes XO cases.
[1] One fetus showed abnormal 45/XY. Other fetus showed breaks, translocation, and fragments. Both from therapeutic abortions.
[2] Translocation of Pair 1.
[3] Includes one fetus from therapeutic abortion who received radiation during pregnancy and had variable/XY chromosome complement.
[4] Fetus from therapeutic abortion.
[5] Triploid in diploid mosaic XXY/XY.
[6] XXY/XY mosaic.
[1] Monosomy.
From *Human Developmental Genetics*, by David Yi-Yung Hsai. Copyright © 1968 Year Book Medical Publishers, Inc. Used by permission.

spontaneous abortions. In the former, there were four abnormalities in 150 cases, or an incidence of 2.6 per cent. In the latter, there were 73 abnormalities in 264 cases, or an incidence of 27.6 per cent. Thus there is a tenfold increase of chromosomal abnormalities in spontaneous as compared with therapeutic abortions.

The abnormalities seen in the therapeutic abortions were the following:

1. A case with 45 chromosomes where two acrocentric chromosomes of group D were missing, and one large metacentric chromosome, similar to chromosome number 3 existed as an extra.
2. A case showing breaks, translocations, and fragmentation in 20 per cent of the chromosomes.
3. A case of tetraploidy with XY complement.
4. A case of variable 45–47/XY chromosome complement with the mother received x-radiation during early pregnancy.

Except for the fourth case, the other three may well represent random abnormalities which might be found in any survey of normal conceptuses.

The abnormalities seen in spontaneous abortions could be divided into five major groups:

1. In thirty-nine cases (47%), there was a disturbance involving one of the autosomes. In one instance, there was a translocation and, in another there was a monosomy, both involving chromosome 1. In all other thirty-seven instances, there was a trisomy involving groups A, B, C, D, E, and G. As with liveborn infants, the abnormality was more common in chromosomes of group G and E and, to a lesser extent, in chromosomes of group D than in groups A, B, and C.
2. In eighteen cases (22%), there were 45 chromosomes with an XO constitution. There was no instances of 47 chromosomes with an XXY constitution.
3. In nineteen cases (23%), there was a triploidy. Of these, 14 cases had 69 chromosomes with an XXY constitution. Penrose and Delhanty have suggested that, in this situation, one set of 23 chromosomes was derived from a sperm and two sets from an ovum. The cause of the triploidy is fertilization of an unreduced ovum which has undergone the first, but not the second meiotic division. In the other triploidy, three had an XXX constitution, one had an XYY constitution, and the constitution of the remaining case is unknown.
4. In four cases (5%), there was a tetraploidy. Of these, two had an XXYY and two had an XYYY constitution.
5. There were three other cases (3%) involving miscellaneous disturbances which has been discussed under therapeutic abortions. Thus, about half of all of the chromosomal abnormalities seen in spontaneous abortions involve a trisomy of one of the autosomes, about a quarter involving either triploidy or tetraploidy, and the remaining quarter involve an XO constitution. It is of interest to note that if one excludes triploids and tetraploids, the ratio of the autosomal to sex chromosome abnormalities in abortions is 2 : 1 compared with 1 : 1 in survivors at birth. This is further evidence of the more lethal character of autosomal trisomy.

BIOCHEMICAL GENETICS

Pediatric literature and clinical experience contain abundant indications of physiological immaturity in newborn infants, especially in those born prior to term. Functional immaturity of brain, liver, kidney, immune response, and blood coagulation are among the well-recognized problems in the newborn period. Exaggerations of these usually mild, transitory phenomena may contribute to grave disorders and possibly prove to be fatal.

From studies on the chemical embryology of the species and from limited similar work with human material, characteristic patterns of chemical differentiation are emerging. Extension of such observations may be expected to permit the establishment of new metabolic parameters by which to assess the developing fetus and infant. Also, studies of certain enzyme systems in the young indicate different levels of activity than are exhibited by the healthy adult. Generally, there appears to be a "deficit" in enzyme activity in the newborn which is more marked in the premature infant and which rises to "normal" levels within a short period. However, in a few instances, the neonate shows higher activity of the particular enzyme, again adjusting in time to the normal pattern of childhood and adult life.

We can illustrate this by examining the phenylalanine-tyrosine system. In mammalian tissues, the breakdown of phenylalanine to acetoacetate is accomplished by a series of six successive steps:

1. Phenylalanine is first converted to tyrosine in the presence of the enzyme system phenylalanine hydroxylase.
2. Tyrosine is next converted to p-hydroxyphenylpyruvic acid in the presence of the enzyme tyrosine transaminase.
3. The p-hydroxyphenylpyruvate is in turn converted to homogentisic acid in one step in the presence of the enzyme p-hydroxyphenylpyruvic oxidase.
4. Homogentisic acid is then converted to maleylacetoacetate in the presence of the enzyme homogentisic oxidase.
5. Maleylacetoacetate is then converted to fumarylacetoacetate by a cisand trans-isomerase, maleylacetoacetic acid isomerase.
6. The fumarylacetoacetate is rapidly split into its component parts by fumarylacetoacetic acid hydrolase, and the pieces enter the general metabolic reaction of the body.

In 1939, Levine and his co-workers first suggested that a derangement of this system may exist in premature infants. They demonstrated excessive amounts of tyrosine and p-hydroxphenyl-pyruvate in the urine of patients receiving a high-protein diet. This effect could be reversed by the administration of ascorbic acid. These findings suggested that metabolic defects existed in these children, involving, first, the conversion of tyrosine to p-hydroxphenylpyruvate and, second, the conversion of p-hydroxy-phenylpyruvate to homogenetisic acid. Subsequently, Kretchmer and his associates showed that there might also be a defect in the conversion of phenylalanine to tyrosine in the newborn infant.

The enzymatic conversion of phenylalanine to tyrosine involves two reactions:

1. Phenylalanine + tetrahydropteridine + $O_2 \rightarrow$ tyrosine + quinoid dihydropteridine + H_2O
2. Quinoid dihydropteridine + NADPH + $H^+ \rightarrow$ Tetrahydropteridine + $NADP^+$

The first reaction is catalized by a labile enzyme which is purified from rat liver and is generally referred to as phenylalanine hydroxylase. The second reaction is catalyzed by a stable enzyme which is purified from sheep liver and is frequently referred to as dihydropteridine reductase. The role of the tetrahydropteridine can be filled by certain synthetic compounds, such as 2-amino-4-hydroxy-6-methyl tetrahydropteridine, or by a naturally occurring cofactor, which has recently been identified as dihydrobiopterin. Thus, one could view the overall reaction of the phenylalanine hydroxylase system as:

$$\text{Phenylalanine} + O_2 + \text{NADPH} + H^+ \xrightarrow[\text{pteridine}]{\text{tetraphydro-}} \text{tyrosine} + H_2O + \text{NADP}$$

In 1959, Kenney and Kretchmer reported that the overall reaction of phenylalanine hydroxylase was virtually absent in the livers of both human premature infants and rats less than twenty-four hours old. In experiments with these rats, they found that supplementing these inactive rat liver fractions with adult rat liver fractions with the labile enzyme would restore normal phenylalanine hydroxylase activity. On the other hand, the activity of fetal rat liver was not affected by the addition of 2-amino-4-hydroxyl-6,7 dimethyl tetrahydropteridine. This suggested that the immaturity was due to a deficiency of the hydroxylase enzyme and not due to a cofactor deficiency.

More recently, Brenneman and Kaufman have reported data which are at variance with those of Kenney and his co-workers. While they noted that the activity of phenylalanine hydroxylase in newborn rats at zero to two hours is significantly lower than the average for adult males and females, the addition of cofactor restored normal adult activity from the time of birth. Detailed studies have shown that the newborn rat has a deficiency of dihydropteridine reductase, but this enzyme reaches adult levels within the first twenty-four hours of life. Finally, by *in vivo* studies measuring the conversion of phenylalanine ^{14}C to tyrosine ^{14}C in three-hour-old rats, they were unable to distinguish any difference between young and adult rats. These findings lead us to believe that the immaturity of phenylalanine hydroxylase seen in newborns represents a deficiency of the coenzyme dihydropteridine reductase rather than of the enzyme phenylalanine hydroxylase.

The widespread screening for phenylketonuria among newborn infants has led to the recognition that an elevation of phenylalanine in the blood can be caused by a variety of factors.

1. *"Classic" Phenylketonuria*

This is the syndrome which was first described by the Norwegian physician and biochemist Ashbörn Fölling in 1934. This hereditary metabolic disease is transmitted by an autosomal recessive gene, and the heterozygous carriers show an excessive retention of L-phenylalanine after an oral load. Most workers are willing to accept the diagnosis of "classic" phenylketonuria if the serum phenylalanine level rises from a normal level of 2 mg/100 ml at birth to a markedly elevated level of 20 mg/100 ml or higher by the end of the first week of life. Further evidence would be a normal serum tyrosine level of 2 mg/100 ml or lower, the presence of o-hydroxyphenylacetic acid in the urine, and, later, the appearance of phenylketones in the urine. Liver biopsy will reveal an absence of phenylalanine hydroxylase activity both with and without the addition of the pteridine cofactor.

2. *Hyperphenylalaninemia Associated with Hypertyrosinemia*

By far the most common cause of hyperphenylalaninemia during early infancy is its association with hypertyrosinemia. Human infants, particularly those with low birth weight, have a functional deficiency of both tyrosine transaminase and p-hydroxyphenylpyruvic acid oxidase. This results in a marked accumulation of tyrosine and a moderate increase of phenylalanine. When the serum phenylalanine and tyrosine levels are plotted against each other, it is apparent that although excessive tyrosine levels do not always result in an increase in the phenylalanine levels, almost every infant with a high phenylalanine level has an elevated tyrosine level.

3. *Hyperphenylalaninemia Associated with Phenylalanine Transaminase Deficiency*

Recently, two reports have appeared describing hyperphenylalaninemia with an absence of phenylketones in the urine. In one of the patients, there is an elevated serum level of phenylalanine and a typical phenylketonuric phenylalanine tolerance test; however, there is an absence of increased concentrations of phenylketones, ortho- and parahydroxyphenyl acids, and phenolic acids, and there is no abnormality in urinary excretion of 5-hydroxyindoleacetic acid or indole-3-acetic acid. These findings would suggest that the patient not only has a deficiency of phenylalanine hydroxylase but also one of phenylalanine transaminase. The combination is probably responsible for the increased levels of serum phenylalanine seen in this infant.

4. *Persistent Hyperphenylalaninemia*

During the past two years, we have seen a number of examples of more persistent hyperphenylalaninemia without phenylketonuria. Most of these patients present a serum phenylalanine level of 20 to 40 mg/100 ml in the first few weeks of life. Subsequently, their level falls to between 5 and 10 mg/100 ml and stays there on a regular diet. They show an elevated phenylalanine tolerance curve, and liver biopsies show a decrease but not an absence of phenylalanine hydroxylase. After excluding hyperphenylalaninemia associated with hypertyrosinemia, this form of persistent hyperphenylalaninemia accounts for about 20 per cent of the infants found to have elevated phenylalanine levels through screening.

5. *Transient Hyperphenylalaninemia*

We have recently encountered an instance of transient hyperphenylalaninemia as a result of a screening program. This baby, one of twins, was found to have an elevated serum phenylalanine level in the range of 20 mg/100 ml and was treated with a low phenylalanine diet. At about six months, his dietary intake of phenylalanine was gradually increased, and eventually he was placed on a regular diet. Subsequently he and his parents were found to have normal phenylalanine tolerance curves. Another such patient has been described in Boston.

With so many different causes for hyperphenylalaninemia during early infancy, it would be difficult on a clinical basis to separate those caused by a transient immaturity of the phenylalanine hydroxylase system from the other possible causes. In fact, it is very unlikely that one would encounter a physiologic immaturity of the phenylalanine hydroxylase system without also encountering a similar deficiency in the tyrosine transaminase and p-hydroxyphenylpyruvic acid oxidase systems, since the latter two tend to show immaturity both more frequently and to a greater degree.

While the therapeutic value of the low-phenylalanine diet remains to be proved by means of a control study, we believe that for the present any patient found to have a plasma phenylalanine level above 20 mg/100 ml and a normal tyrosine level should be put on a low-phenylalanine diet.

At three, six, and twelve months of age, the child is taken off the low-phenylalanine diet and placed on a regular diet until a rise of plasma phenylalanine to 20 mg/100 ml is achieved.

From the practical point of view, it is unlikely that the elevations of plasma phenylalanine level associated with an immaturity of phenylalanine hydroxylase, tyrosine transaminase, or p-hydrophenylpyruvic acid oxidase will ever rise to such levels and, in most instances, no treatment is indicated. Recently, the suggestion has been made that folic acid, which is the nearest chemical compound to tetrahydropteridine, might be administered to normal birth weight and low birth weight infants to see if this would in any way alter the plasma phenylalanine and tyrosine levels.

IMMUNOGENETICS

We can illustrate the maternal and fetal interraction at an immunological level by the recent work on the prevention of hemolytic disease. About one in every two hundred newborn infants will be affected by hemolytic disease, and, of these, about one in ten will die either from hydrops fetalis or kernicterus. In the usual situation, the disease occurs when the mother is Rh negative and the father is Rh positive. If the father is homozygous, all of their children will be RH positive. If the father is heterozygous, there is a 50 per cent chance that their children will be Rh negative. Normally, the placenta would be expected to prevent the infant cells, which are Rh positive, from entering the maternal circulation, which is Rh negative. However, this transfer occurs in certain instances, and, in these situations, the mothers develop anti-Rh antibodies, and the infants develop erythroblastosis fetalis. Until recently, treatment of Rh hemolytic disease consisted solely of performing exchange transfusions after the birth of the infant, or, at most, doing intrauterine exchange transfusions following spectrophotometric analysis of amniotic fluid.

In 1961, Finn and his co-workers in Liverpool suggested that one way to prevent Rh immunization of mothers would be to destroy the fetal cells as rapidly as possible after they enter the circulation of the mother. This could be done by giving fraction II gamma globulin made from a pool of high-titer incomplete anti-Rh donor plasma (RhoGAM, Ortho Diagnostics, Raritan, N.J.). When 5 ml of this Rh immunoglobulin is administered intramuscularly, it will give a maximum artificial antiglobulin titer of 1 : 128. Similarly, 1 ml will give a titer of 1 : 32, and 0.1 ml will give a titer of 1 : 2.

Controlled trials on the use of Rh immunoglobulin are currently being carried out in New York, California, Edinburgh, Liverpool, and Freiburg. The combined results of these trials up to the summer of 1967 are summarized in Table 2-V. It can be seen that none of the 582 protected mothers had become actively immunized, while sixty-five of the 537 control mothers (12.1%) had developed immune antibodies to Rh. At subsequent delivery, there were no sensitized babies in the seventy-two protected women, whereas twelve of the sixty-four control mothers (20.3%) had an Rh-affected infant. These data show clearly that anti-Rh immunoglobulin preparation is effective in preventing Rh sensitization among women who have not been immunized. However, conventional methods of treatment will still have to be used for women who have already been sensitized to Rh antibodies.

TABLE 2-V
CLINICAL TRIAL OF PREVENTION OF IMMUNIZATION OF RH-NEGATIVE MOTHERS AT RISK*

	By 6 Months After Delivery		*At Subsequent Delivery*	
	Rh Ig G Treated	*Controls*	*Rh Ig G Treated*	*Controls*
No. tested	582	537	72	64
No. sensitized	0	65	0	13
Percentage sensitized	0%	12.1%	0%	20.3%

* Combined results as tabulated in Freda, V.J., Gorman, J.G., and Pollack, W.: Suppression of the primary Rh immune response with passive Rh Ig G immunoglobulin. *New Engl J Med 277:* 1022, 1967.

IATROGENIC FACTORS

Tetratology is a comparatively new discipline. Until recently, most physicians assumed that all congenital malformations were genetic in origin. Pedigrees which did not fit classical mendelian principles were either written off with such vague explanations as variability of gene expression or cited as examples of fresh mutations. In 1929, Murphy and Golstein first noted that pelvic radiotherapy during pregnancy is extremely injurious to the fetus. In a survey covering 106 women who had received postconception roentgen ray or radium treatments, somewhat more than one third of the children had gross mental and physical abnormalities with microcephalus being the most prominent. Twelve years later, Gregg, an ophthalmologist in Australia, established a definite relationship between rubella during pregnancy and the occurrence of congenital cataracts and other defects. More recently, Lenz and others have shown that if mothers took thalidomide in early pregnancy, their offspring would have amelia, phocomelia, and absence of the radius and radial digits. These observations provided definite proof that congenital malformations could be caused by nongenetic pathologic modification of normal embryos. We shall discuss the latter condition in somewhat greater detail.

Phocomelia, which means "seal extremities" (the word comes from two Greek words, *phokos* meaning "seal" and *melos* meaning "extremities"), is an extremely rare congenital anomaly. The artist Goya recorded a case in one of his paintings which now hangs in the Louvre. During the early years of the present century, one or two additional cases, usually affecting only one limb, were reported in the literature. In 1960, Kosenow and Pfeiffer presented an exhibit at the German Pediatric Meeting showing two examples of phocomelia where there were no known genetic factors, blood group incompatibilities, or chromosomal abnormalities. Because of the rarity of the syndrome little attention was paid to the exhibit. Yet, as shown in Table 2-VI, the incidence of phocomelia in 1960 reached

TABLE 2-VI
INCIDENCE OF PHOCOMELIA IN VARIOUS GERMAN UNIVERSITY PEDIATRIC CLINICS*

	1949–1959	*1959*	*1960*	*1961*	*1959–1961*
Bonn		2	19	50	71
Bremen			4	20	24
Frankfurt		1	4	11	16
Göttingen		3	1	10	14
Hamburg		1	30	154	185
Heidelberg		2	5	9	16
Kiel		2	4	26	32
München	3	2	14	44	60
Münster	4/yr.	3	27	96	126

* From Taussig, H.B.: A study of the German outbreak of phocomelia. *JAMA, 180:* 1106, 1962.

epidemic proportions in Germany and, shortly after the exhibit, it was found that the outbreak was related to the use of thalidomide by mothers during pregnancy.

The most common and striking abnormalities caused by thalidomide are reduction deformities of the limbs. The effect on the upper limb is always to inhibit development of a part or the whole. There may be hypogenesis of the thumb, or the thumb and the index finger. The remaining fingers may be deformed and there is often fusion of two adjacent fingers, resulting in syndactyly. If the radius is absent, the thumb is always absent, and the index finger often is absent as well. There may be complete absence or deformity of the radius, humerus, or ulna. In the absence of all long bones, there may still be one or two digits (presumably ulnar digits) protruding from the shoulder. Finally, there may be complete absence of the arm.

The effect on the lower limb is generally similar, but differs in some important respects. The feet are usually normal in structure, whereas the femur, tibia, and fibula may be deformed or absent. When there is no normal femoral head, there is no normal acetabulum. Unlike the fingers, which are often absent or fused, the toes are often widely spaced and supernumerary.

Other anomalies seen in the thalidomide syndrome are hypoplasia and absence of the pinna in the ear, microphthalmia with coloboma of the iris and retina, congenital heart disease (principally tetralogy of Fallot and ventricular septal defect), obstructive lesions of the intestine, and renal agenesis and hypoplasia. Most children with the thalidomide syndrome show normal mental development.

Thalidomide is a synthetic drug with no direct toxic effects on animals or man. Because of its wide margin of safety, it was sold without a prescription as a sleeping tablet, sedative, or tranquilizer. On November 20, 1961, Lenz from Hamburg, Germany, reported to the Dusseldorf Pediatric Meeting that he suspected a specific drug to be the cause of the phocomelia syndrome and that he had warned the company manufacturing the drug to withdraw it from the market. Almost simultaneously, McBride from New South Wales in Australia sent a letter to the *Lancet* stating that the incidence of multiple severe abnormalities in babies delivered of mothers who were given thalidomide was 20 per cent as compared with the usual 1.5 per cent. Finally, Speirs from Stirlingshire in Wales found that at least eight of ten mothers who gave birth to babies with major limb defects had received thalidomide in the early weeks of pregnancy. By the time thalidomide was withdrawn by the manufacturers, the phocomelia syndrome had been seen in Sweden, Belgium, Switzerland, Lebanon, Israel, Peru, Canada, Brazil, the Netherlands, the United States, and Japan. There was a close correlation between the frequency of affected births and the supply of thalidomide one year earlier.

According to Lenz, the embryo is most sensitive to thalidomide between the thirty-fourth and fiftieth days after the last menstrual period. Ingestion of the drug early in pregnancy causes abnormalities of the ear and thumbs, followed by reduction deformities of the upper limbs and then the lower limbs, and finally by gastrointestinal and cardiac anomalies. Lenz has presented convincing evidence that these malformations will appear in many newborn infants born of mothers who received thalidomide between the third and eighth weeks after conception and that sometimes a single dose of 200 mg given at a crucial point can cause phocomelia.

Despite the obvious association between thalidomide and congenital malformations in man, it has been difficult to reproduce a similar teratogenic effect in experimental animals. This has greatly handicapped the study on the pathogenesis of the thalidomide effect. It appears that thalidomide influences the metabolism at the cellular level. Roath, Elves, and Israels have shown that the addition of the drug to leukocyte cultures at concentrations of 25 to 50/ml inhibits cell multiplication. It has been suggested that thalidomide might act as an antagonist of glutamine or glutamic acid. However, definitive proof of this either in *in vivo* or in *in vitro* systems has not come forth. Similarly, the theory that thalidomide might act as a riboflavin antagonist or operate in a preexisting riboflavin deficient environment remains to be proved.

REFERENCES

1. Brenneman, A.R., and Kaufman, S.: Characteristics of the hepatic phenylalanine-hydroxylating system in newborn rats. *Gen Biol Chem, 240:*3617, 1965.
2. Carr, D.H.: Cytogenetics of abortions. In Benirschke, K. (Ed.): *Comparative Aspects of Reproductive Failure.* New York, Springer-Verlag, 1967.
3. Clarke, C.A., *et al.:* Prevention of Rh-haemolytic disease: result of clinical trial and combined study from centers in England and Baltimore. *Br Med J. 2:*907, 1966.
4. Hsia, D.Y.-Y: *Human Development Genetics.* Chicago, Year Book Pub., 1968.
5. Justice, P.; O'Flynn, M.E., and Hsio, D.Y.-Y.: Phenylalanine hydroxylase activity in hyperphenylalanemia. *Lancet, 1:*928, 1967.
6. Lenz, W.: Malformations caused by drugs in pregnancy. *Am J Dis Child, 112:*99, 1966.
7. Penrose, L.S.: Genetics of growth and development of the fetus. In Penrose, L.S. (Ed.): *Recent Advances in Human Genetics.* Boston, Little Brown, pp. 56–75.

Chapter 3

Some Basic Factors Involved in the Production of Neuromuscular Block by Various Quaternary Ammonium Compounds

William L. Nastuk

Neuromuscular transmission involves a series of events which can be summarized as follows: motor neuron action potential → depolarization of neuronal terminals → release of acetylcholine (ach) into synaptic cleft → diffusion of ach to muscle postjunctional membrane (PJM) → combination of ach with PJM receptor sites → increase in PJM permeability to Na^+ and K^+ → depolarization of the PJM (endplate potential) → initiation of propagated muscle action potential → excitation-contraction coupling → muscle tension development. These steps themselves depend on many processes and modifying factors some of which are reviewed in Figure 3–1.

In recent years knowledge of the fundamentals of the physiology and pharmacology of neuromuscular and synaptic transmission has increased remarkably. Experimenters who work in this area rely on a wide variety of experimental approaches, but frequently their investigations are carried out on isolated tissues *in vitro* and single fiber techniques are often employed. The model in Figure 3–2 illustrates several electrophysiological techniques commonly utilized in our laboratory. In an isolated muscle, membrane potentials of single fibers are recorded with an intracellular electrode (*lower right*). The composition of the fluid bathing a single neuromuscular junction is altered by microperfusing this zone with appropriate solutions (*left*). Drugs are applied at high speed to a discrete area of membrane by means of iontophoresis (*upper right*). More details of these and related methods can be found in reference 1.

Various *in vitro* techniques such as those mentioned above offer powerful advantages in the study of neuromuscular transmission. For example: (1) membrane potential and membrane conductance records of a relatively uncomplicated nature can be obtained from single fibers; (2) membrane potentials can be controlled and manipulated by voltage clamp-

This work was supported by National Institutes of Health Research Grant 5RO1–NB–04988.

Presynaptic Elements
- Acetylcholine
  - Synthesis
  - Storage
  - Mobilization
  - Release
    - Membrane Depolarization
      - (via AP or Electrotonically)
    - Ca, Mg, Botulinum Toxin, etc.
- Extent of Terminal Arborization

Synaptic Cleft
- Diffusion of Ach

Postsynaptic Elements
- Extent of PJM Area
- Reaction of PJM Receptor Sites to Ach
  - Desensitization Mechanism
- RP of Muscle Fiber
  - Ionic Composition of Extra- and Intracellular Fluids
- Excitability of conductile system
- Hydrolysis of Ach

Contractile Elements
- Excitation-Contraction Coupling
- Performance of the Acto-Myosin System

Figure 3–1. List of factors important in the functioning of both neuromuscular transmission and the contractile elements of muscle fibers. These factors show the possibilities where interruption of function may occur and produce clinical signs of muscular weakness.

ing; (3) drugs can be applied to chemoreceptive regions and can be maintained at more accurately known concentrations.

Many important and needed experiments concerning clinical aspects of neuromuscular transmission are conducted on whole animals or on patients undergoing surgical procedures. In the latter case the anesthesiologist-experimenter generally administers several drugs or anesthetic agents as a part of routine clinical procedure and some of the agents administered may have action at the neuromuscular junction. Thus these agents can influence the patient's response to a specific blocking agent whose effectiveness or mode of action is being evaluated. In addition,

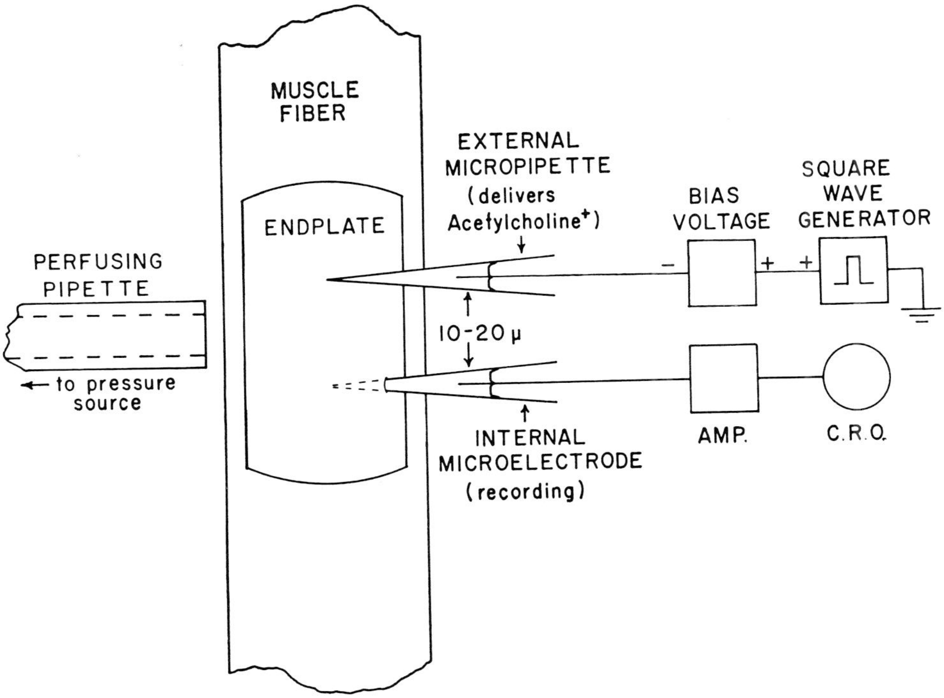

Figure 3–2. Schematic diagram showing the experimental arrangement used when recording changes in postjunctional membrane potential during iontophoretic application of ions, such as acetylcholine+, to the PJM surface. *Left,* Pipette containing solution to be applied to the junctional area by perfusion. *Upper right,* Iontophoresis apparatus. *Lower right,* Intracellular electrode and associated gear for recording transmembrane potentials.

at the neuromuscular junction the concentration of the administered blocking agent changes with time and perhaps with various clinical manipulations. For these and other reasons it is difficult and sometimes impossible to utilize the responses of a patient as means for explaining the fundamental basis of action of a specific agent.

To illustrate the problem of using clinical information in interpreting the action of a drug, let us consider the following. In laboratory experiments, the tension output of an indirectly stimulated isolated nerve muscle preparation can be measured when endrophonium is added to the bathing solution. Endrophonium, a rapid-acting anticholinesterase agent, is an effective antagonist of neuromuscular block produced by d-tubocurarine (Fig. 3–3). However, endrophonium does not effectively antagonize neuromuscular block produced by diethyl ether (Fig. 3–4), or by hexafluorenium (Fig. 3–5). This fact might be regarded as paradoxical because diethyl ether and hexafluorenium have been thought to have

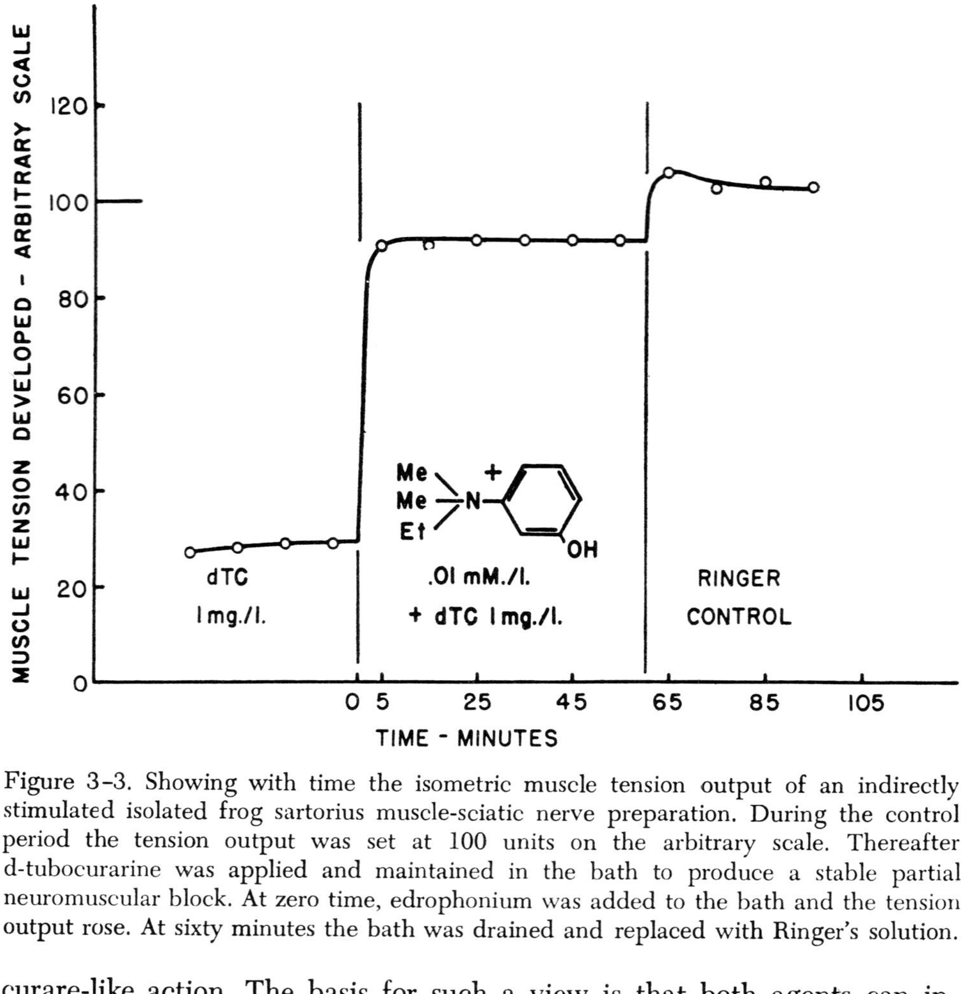

Figure 3-3. Showing with time the isometric muscle tension output of an indirectly stimulated isolated frog sartorius muscle-sciatic nerve preparation. During the control period the tension output was set at 100 units on the arbitrary scale. Thereafter d-tubocurarine was applied and maintained in the bath to produce a stable partial neuromuscular block. At zero time, edrophonium was added to the bath and the tension output rose. At sixty minutes the bath was drained and replaced with Ringer's solution.

curare-like action. The basis for such a view is that both agents can intensify partial neuromuscular block produced by d-tubocurarine. But this appears to be specious reasoning because direct evidence obtained from single fiber *in vitro* experiments indicates that diethyl ether and hexafluorenium do not act as competitive blocking agents as is true for d-tubocurarine.[2,3]

Let us consider another example. It is well known that desensitization block can be produced at the neuromuscular junction by prolonged application of quarternary ammonium compounds which can depolarize the muscle postjunctional membrane (PJM). Typical compounds are carbamylcholine (carb), decamethonium, and succinylcholine. Desensitization block (sometimes called Phase II block) has been thought to have characteristics resembling the block produced by application of d-tubocurarine. However, the evidence supporting this view is indirect and

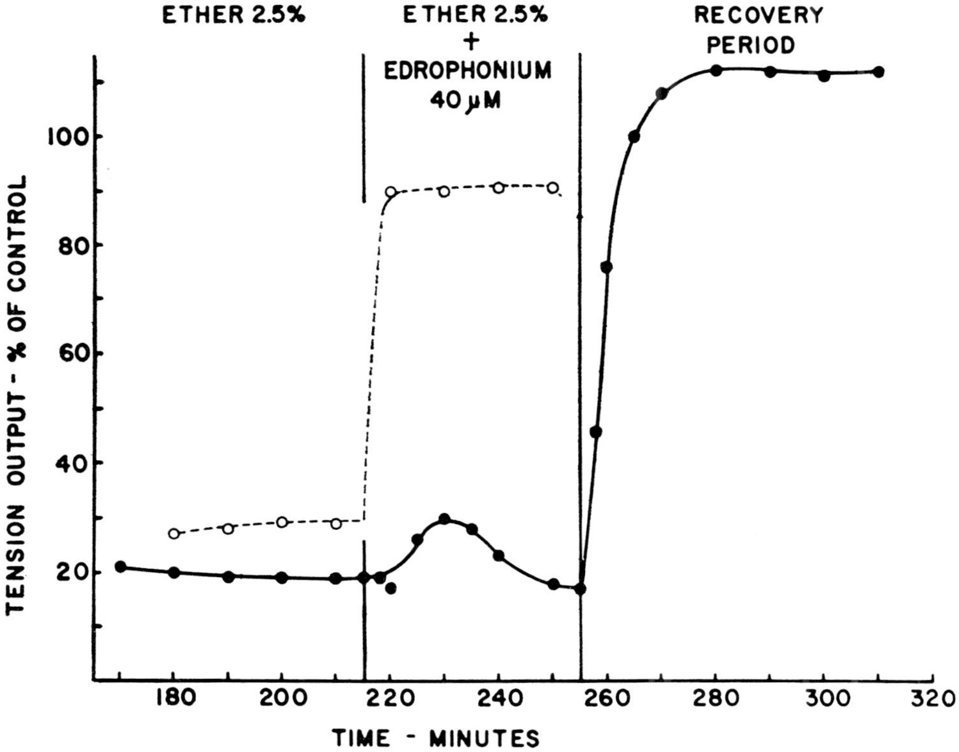

Figure 3–4. Solid circles show fall in isometric tension output of indirectly stimulated isolated frog sartorius muscle at various times after addition of diethylether and diethylether plus edrophonium. Open circles—plot taken from Figure 3–3 to compare the action of edrophonium on a d-tubocurarine-produced neuromuscular block.

complicated. In the laboratory under controlled conditions, one can readily show (Fig. 3–6) that endrophonium does not effectively antagonize a deep desensitization block produced by carb. But it is also possible to show that in nerve-muscle preparations in which the PJM is *not* desensitized by pretreatment with carb, application of endrophonium causes a marked increase in the tension output of indirectly stimulated muscle (Fig. 3–7). One can reasonably suppose that between these two extremes there exists a continuum of intermediate states in which desensitization of the postjunctional membrane ranges from virtually complete to insignificant. From this it is reasonable to infer that as the intensity of PJM desensitization decreases, one may expect to obtain an increased amount of facilitatory action on muscle tension output in response to applied endrophonium.

During recovery of a patient from neuromuscular block produced by application of a depolarizing agent such as succinylcholine, it has in fact

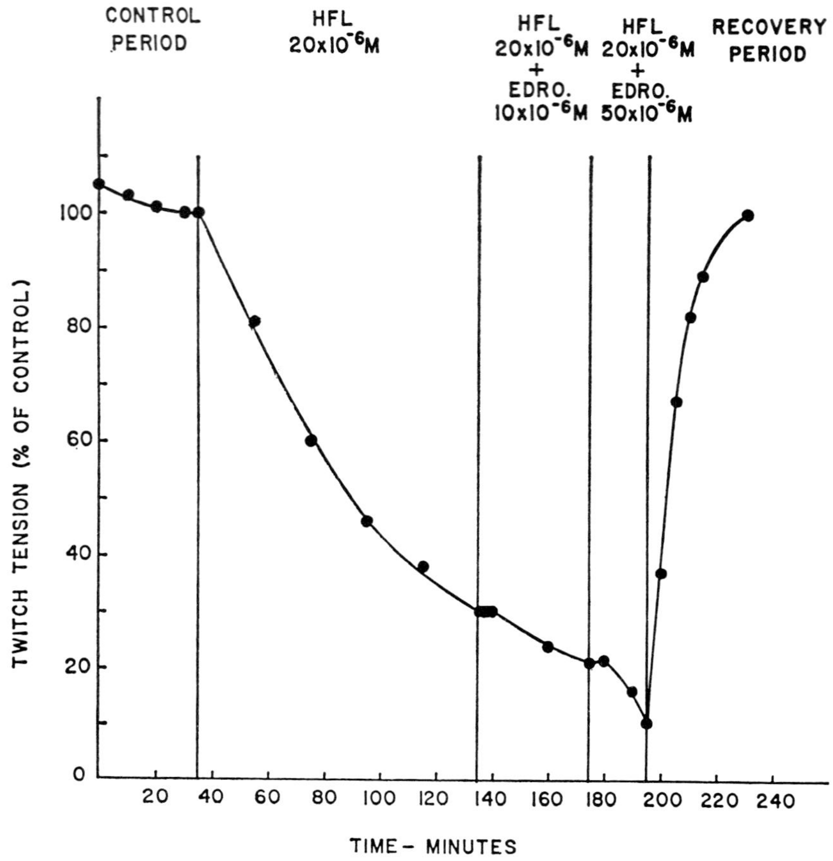

Figure 3–5. Isometric tension output of indirectly stimulated isolated frog sartorius muscle vs. time showing development of neuromuscular block produced by bath-applied hexafluorenium and also showing failure of edrophonium to antagonize this block.

been observed that muscle tension output can be increased by administration of endrophonium. This clinical observation is considered useful because to some it represents one approach to speeding the postoperative recovery of the surgical patient. Whether one accepts this or not, it is certainly clear that such an observation cannot be taken as evidence that desensitization block, as described above, is curare-like.

Many aspects of the neuromuscular block produced by depolarizing quaternary ammonium compounds offer challenging questions to the anesthesiologist, the pharmacologist, and the physiologist. In our laboratory we have been studying some of the fundamental mechanisms involved in the production of neuromuscular block of this kind. If one records the tension output of indirectly stimulated muscle *in vitro,* it can be shown (Fig. 3–8) that application of carb in appropriate concentra-

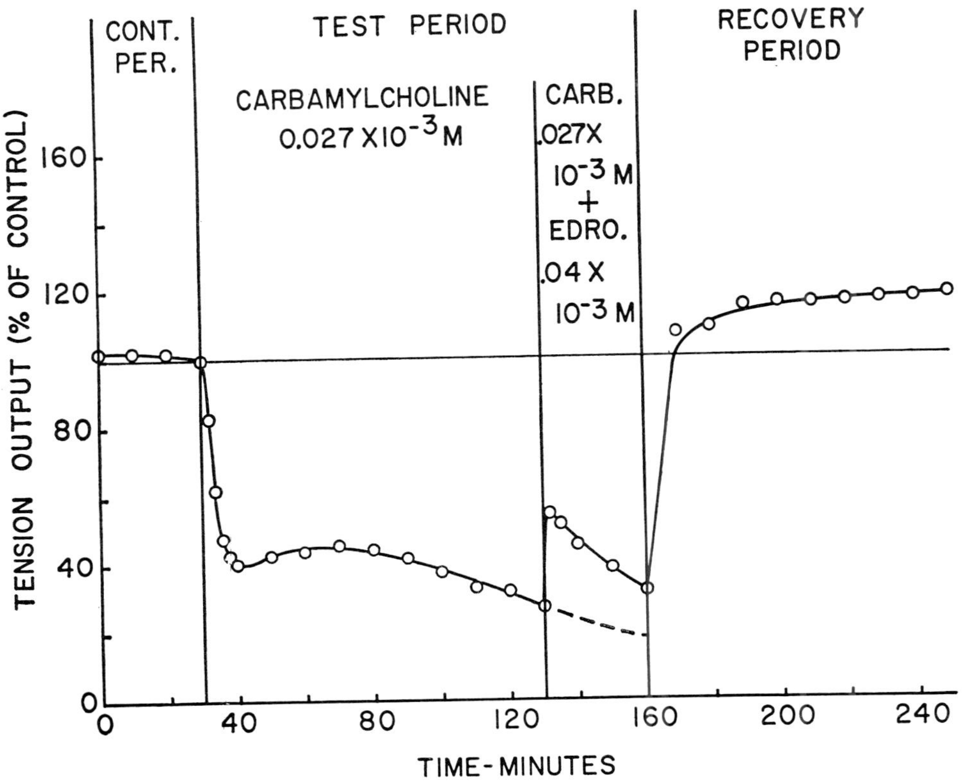

Figure 3–6. Showing the isometric tension output of indirectly stimulated isolated frog sartorius muscle vs. time. Addition of carbamylcholine 0.027 mM produced a "desensitization block" which was not effectively antagonized by addition of edrophonium 0.04 mM.

tion produces a diphasic neuromuscular block.[4,5] Analysis shows that the early phase of block depends largely on the intense depolarization of the PJM which renders the muscle fiber electrically inexcitable. During continuous application of carb membrane repolarization occurs (Fig. 3–8) and, with time, electrical excitability is regained. However, PJM desensitization (which is considered to begin its development from the moment that carb is applied) becomes more and more intense. As a result of this increasing PJM desensitization neuromuscular block reappears. Note (Fig. 3–8) that appreciable PJM depolarization occurs throughout the entire period of carb application. Furthermore, in these experiments it can be seen that following removal of carb, PJM repolarization is accelerated. This change indicates that PJM receptor activation occurred continuously throughout the period of carb application.

If carb is applied at a lower concentration than that used in the pre-

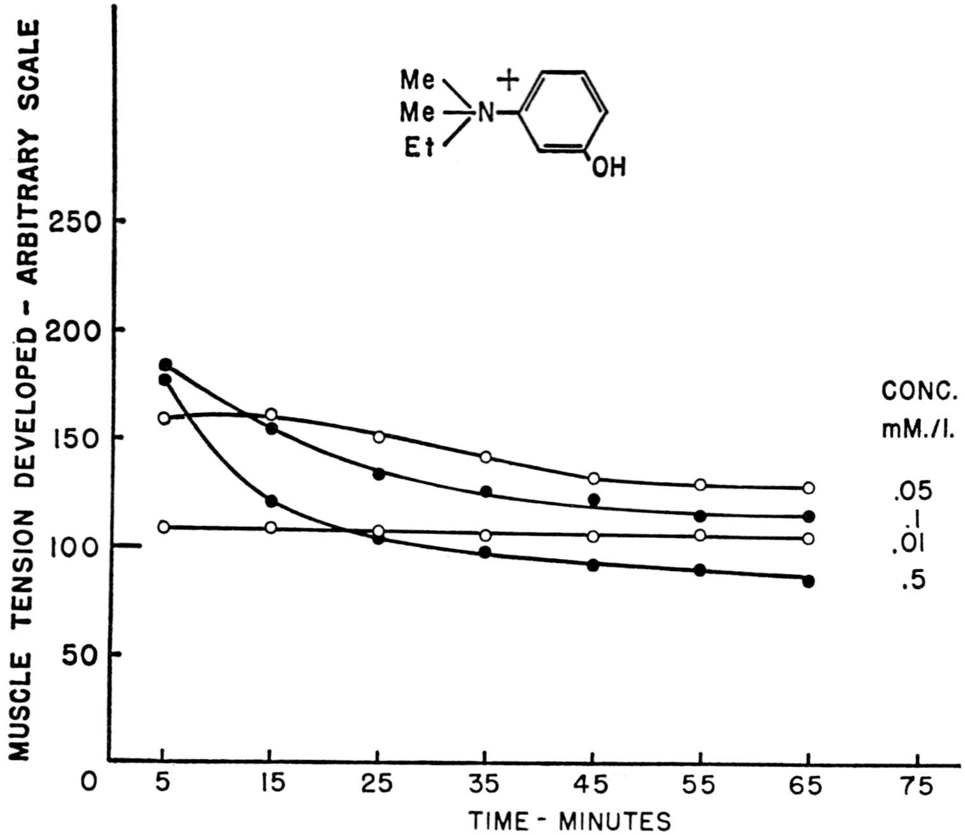

Figure 3–7. Showing vs. time the isometric tension output of indirectly stimulated isolated frog sartorius muscle after bath application of edrophonium in four concentrations. The isometric tension output recorded during the control period was set at 100 on the arbitrary Y-axis scale. Note the failure of edrophonium to potentiate the tension output when this drug is administered in high concentration.

ceding experiments, no neuromuscular block is produced. On the contrary, the tension output is increased above the control value (Fig. 3–9). In contrast when carb is applied in relatively high concentrations, complete neuromuscular block appears early and thereafter it shows no diphasic character. PJM repolarization is prominent under such conditions and the PJM shows a progressive loss of depolarization response to carb (Fig. 3–10).

In general, increasing the concentration of continuously applied carb increases both the extent of the PJM depolarization produced initially and the rate of subsequently developed repolarization (Fig. 3–11). Thus the concentration of the applied electrogenic quaternary ammonium compound represents a powerful variable which governs PJM repolarization and desensitization.

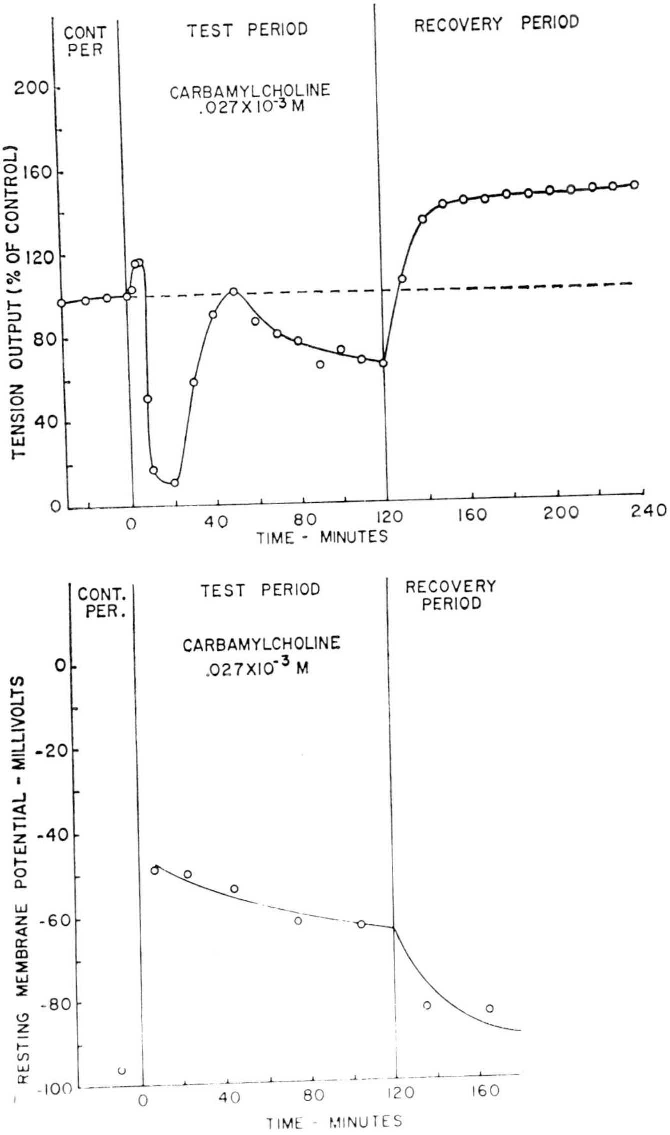

Figure 3–8. *Upper plot,* showing effect of carbamylcholine 0.027 mM on the isometric tension output of indirectly stimulated frog sartorius muscle (average of 6 experiments). *Lower plot,* membrane resting potentials of single muscle fibers measured at the PJM at various times after bath application of carbamylcholine 0.027 mM. Each point represents the average of at least ten fibers. Compare the time course of PJM potential with the tension output as shown above.

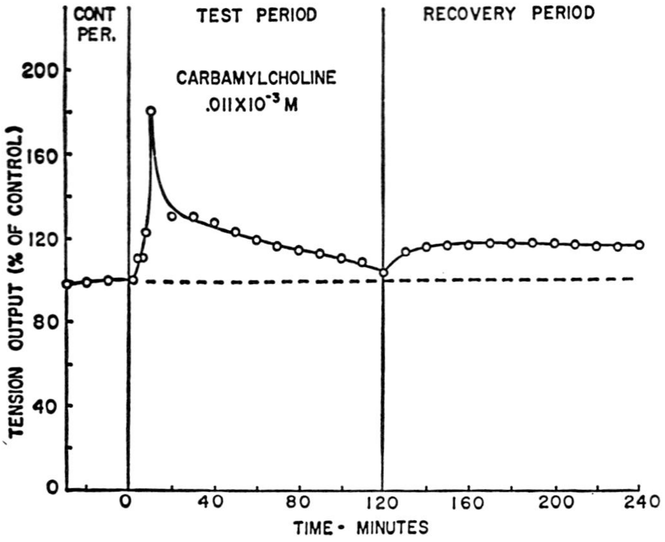

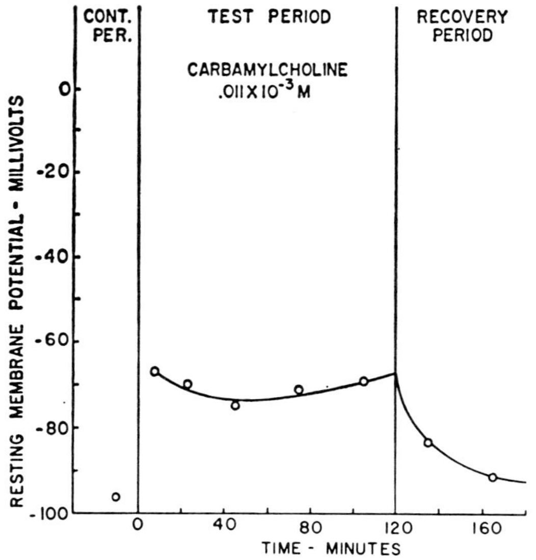

Figure 3–9. *Upper plot,* effect of carbamylcholine 0.011 mM on the isometric tension output of indirectly stimulated frog sartorius muscle (average of 6 experiments). *Lower plot,* membrane resting potential of single muscle fibers measured at the PJM at various times after bath application of carbamylcholine 0.011 mM. Each point equals average of ten or more fibers.

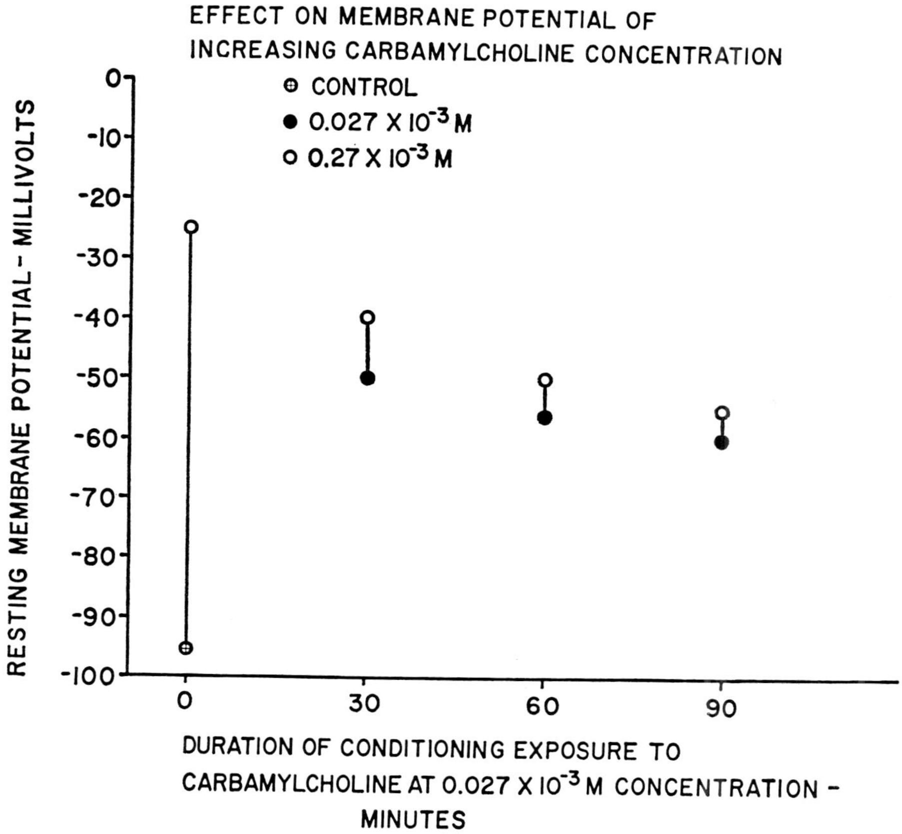

Figure 3–10. Showing the effect on the postjunctional membrane potential of applying carbamylcholine 0.27 mM before and at various times during the continuous application of carbamylcholine 0.027 mM. Note the progressive loss in the depolarizing power of 0.27 mM carb.

It is clear that other factors also influence the rate and extent of PJM desensitization. The chemical structure of the applied quaternary ammonium compound has important influence on PJM desensitization. This fact can be appreciated from the rate and extent of repolarization obtained with acetylcholine (Fig. 3–12) and phenyltrimethylammonium (Fig. 3–13).

Ca^{2+}, one of the ionic constituents in the extracellular fluid, has a most prominent influence in accelerating PJM desensitization. This fact was shown by Manthey,[6] one of the former members of our laboratory. Some typical data which he obtained are shown in Figure 3–14. Nastuk and Parsons[7] have expanded and extended these earlier findings and have included additional variables governing desensitization.

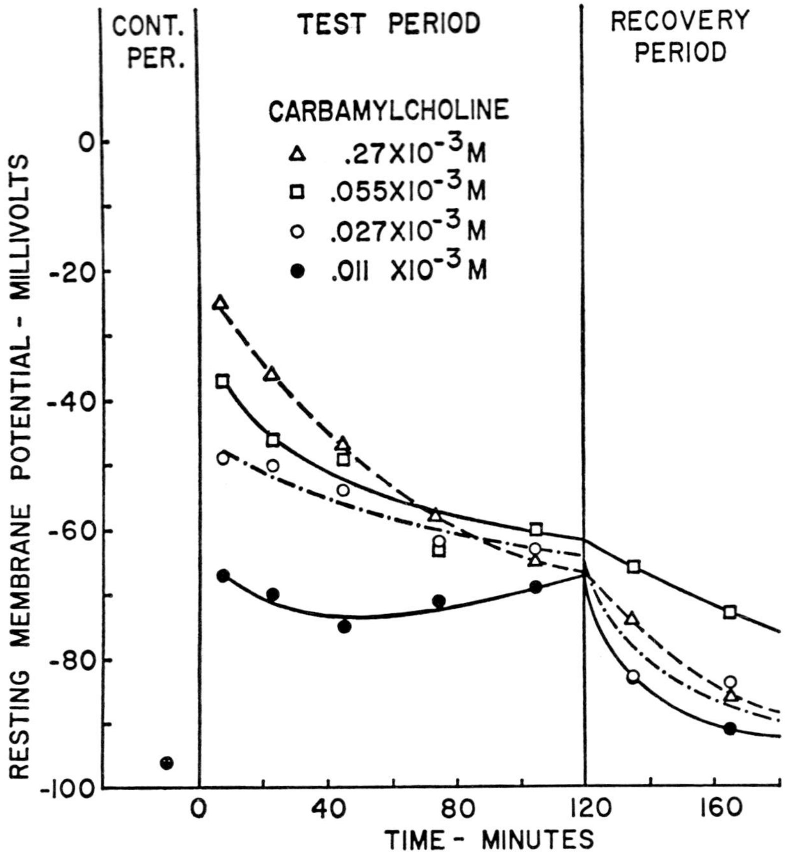

Figure 3–11. Showing postjunctional membrane potentials recorded from single muscle fibers of frog sartorius muscle at various times after bath application of carbamylcholine in a range of concentrations. Each point equals average of ten or more fibers.

Although PJM receptor desensitization (perhaps better called receptor inactivation) has been known and studied for many years, the intimate details involved in this process at the molecular level are not understood. Many as yet unexplained factors must be taken into account in formulating a molecular model for the process. Among the important points are the following: (1) increasing the concentration of extracellular Ca^{2+} accelerates PJM receptor inactivation; (2) it is necessary to apply drugs such as carb in relatively high concentration in order to produce rapid, complete PJM receptor inactivation; (3) the rate of receptor inactivation is relatively slow compared with receptor activation (which generates PJM depolarization).

An early attempt to formulate a model for receptor desensitization was

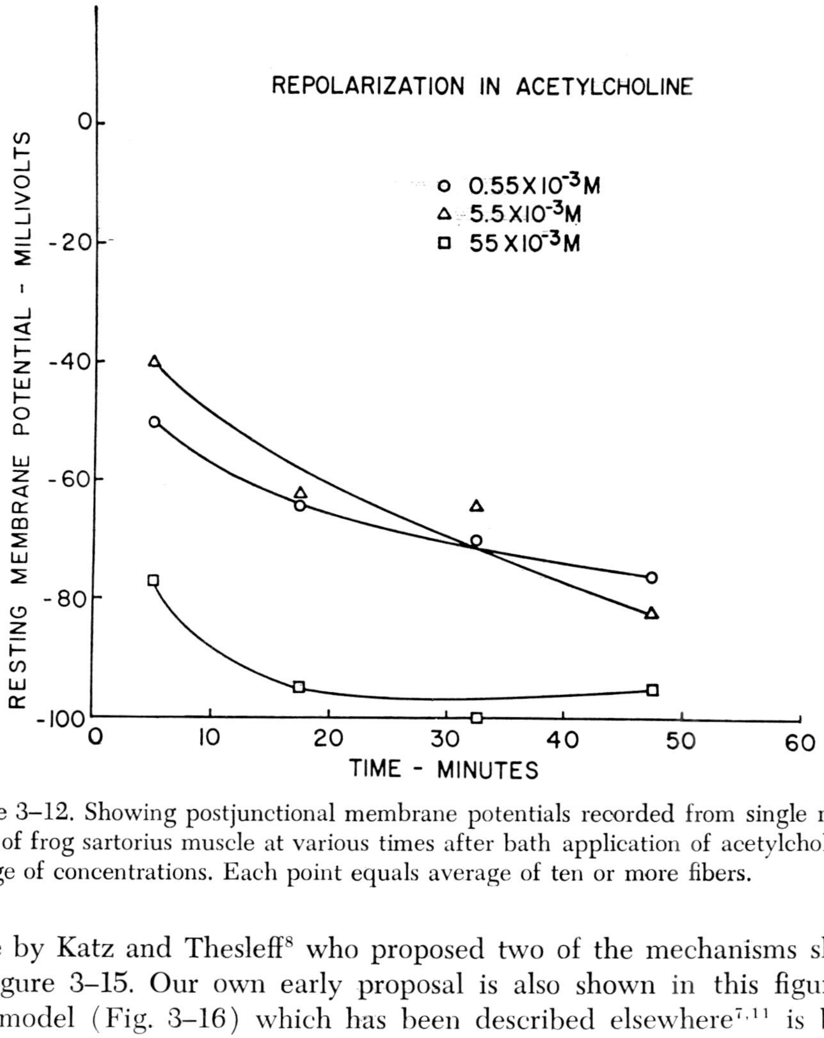

Figure 3–12. Showing postjunctional membrane potentials recorded from single muscle fibers of frog sartorius muscle at various times after bath application of acetylcholine in a range of concentrations. Each point equals average of ten or more fibers.

made by Katz and Thesleff[8] who proposed two of the mechanisms shown in Figure 3–15. Our own early proposal is also shown in this figure. A new model (Fig. 3–16) which has been described elsewhere[7,11] is based on past and more recent work of our laboratory. An important new feature of this model is the idea that PJM receptor desensitization depends on increase in *intracellular* Ca^{2+} concentration and that intracellular Ca^{2+} ions can combine with anionic receptor sites on the inner surface of the PJM. We have obtained indirect histochemical evidence[8,9] which indicates that Ca^{2+} accumulates at the neuromuscular junction of muscle fibers treated with depolarizing quaternary ammonium compounds alone or in combination with caffeine. Figure 3–17, which unfortunately is not reproduced in its original color, shows such junctionally accumulated Ca^{2+} which has been precipitated as the Ca^{2+} salt of the dye alizarin red S.

We are now attempting to test the intracellular action of Ca^{2+} on

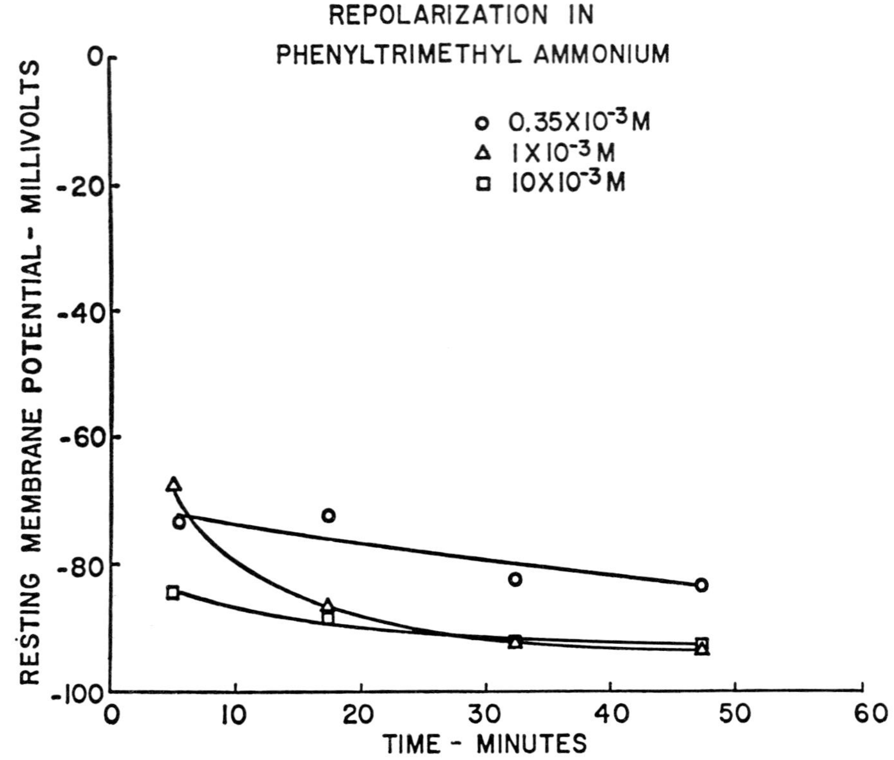

Figure 3–13. Showing postjunctional membrane potentials recorded from single muscle fibers of frog sartorius muscle at various times after bath application of phenyltrimethyl-ammonium in a range of concentrations. Each point equals average of ten or more fibers.

receptor desensitization by means of more direct experiments, and despite some technical difficulties we hope soon to present more definitive evidence on this question. In retrospect, it is interesting to note that numerous experiments have been conducted by many investigators to test the effect of changes in extracellular [Ca^{2+}] on various excitable cells but much less is known of the intracellular action of Ca^{2+} on these tissues, and we have the view that this may be a matter of far-reaching importance. Whatever the value of the proposed model, it seems certain that the understanding of the nature of neuromuscular block at the molecular level must depend on further advances in the knowledge of membrane receptors and the role they play in membrane permeability controlling mechanisms.

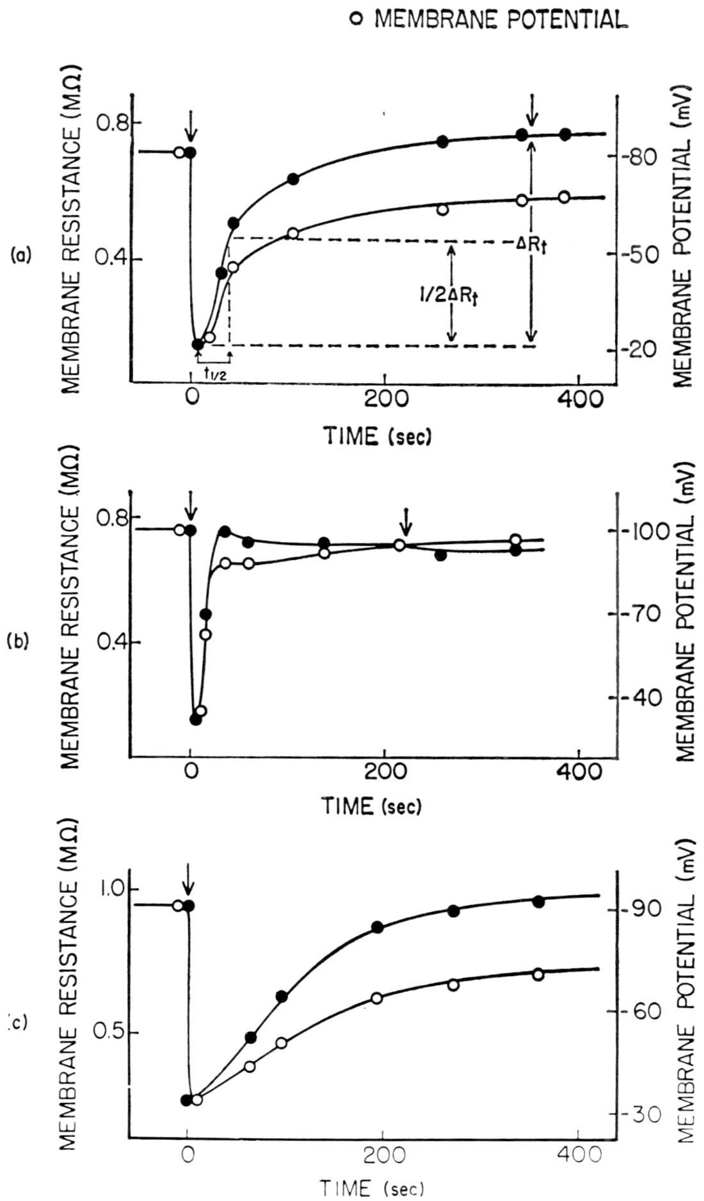

Figure 3–14. Showing time course of changes in effective membrane resistance and membrane potential measured at region of postjunctional membrane. Records (*a*), (*b*), and (*c*) represent measurements from experiments performed on muscle fibers equilibrated in hypertonic Ringer's-sucrose solutions containing 1.8, 10, and 0 mM Ca^{+} respectively. In each case the closed circles represent measurements of effective membrane resistance and the open circles, postjunctional membrane potential. Vertical arrows indicate the onset and end of microperfusion of neuromuscular junction with 0.27 mM carbamylcholine. The method of calculating the half-time of desensitization is indicated in *a*. Note the effect of Ca^{2+} in accelerating desensitization. For further details see Manthey.[6]

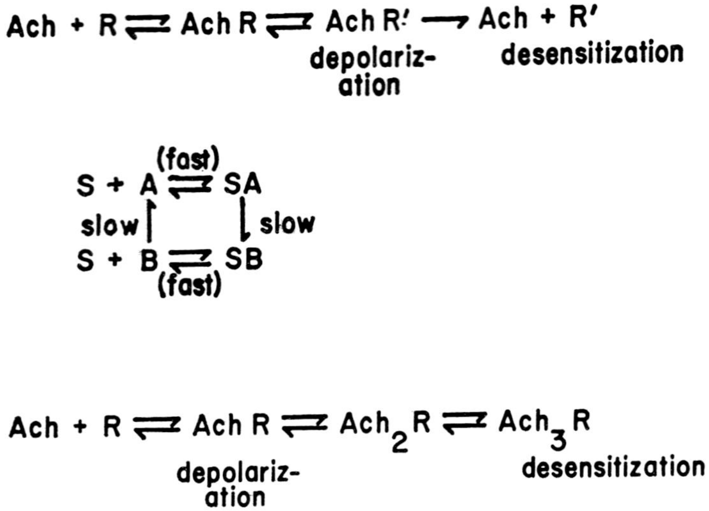

Figure 3–15. Early molecular models for explaining postjunctional receptor desensitization. The upper two are from Katz and Thesleff[8] and the lower is from our own laboratory.[5]

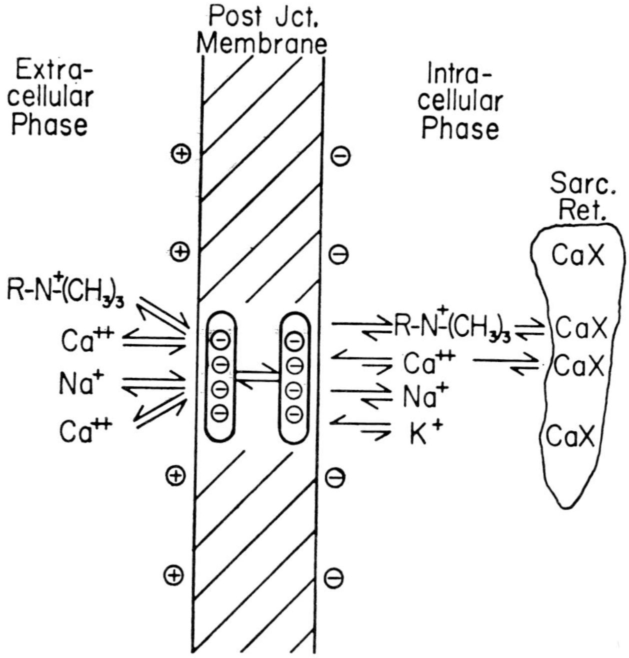

Figure 3–16. Newly proposed molecular model for explaining postjunctional receptor desensitization which takes into account additional factors and includes the speculation that the intracellular $[Ca^{2+}]_i$ concentration is an additional important variable in the receptor desensitization process. In this scheme the distribution of a multivalent ion carrier in the PJM is considered to depend on Ca^{2+} ions which enter into strong combination with anionic sites on the carrier.

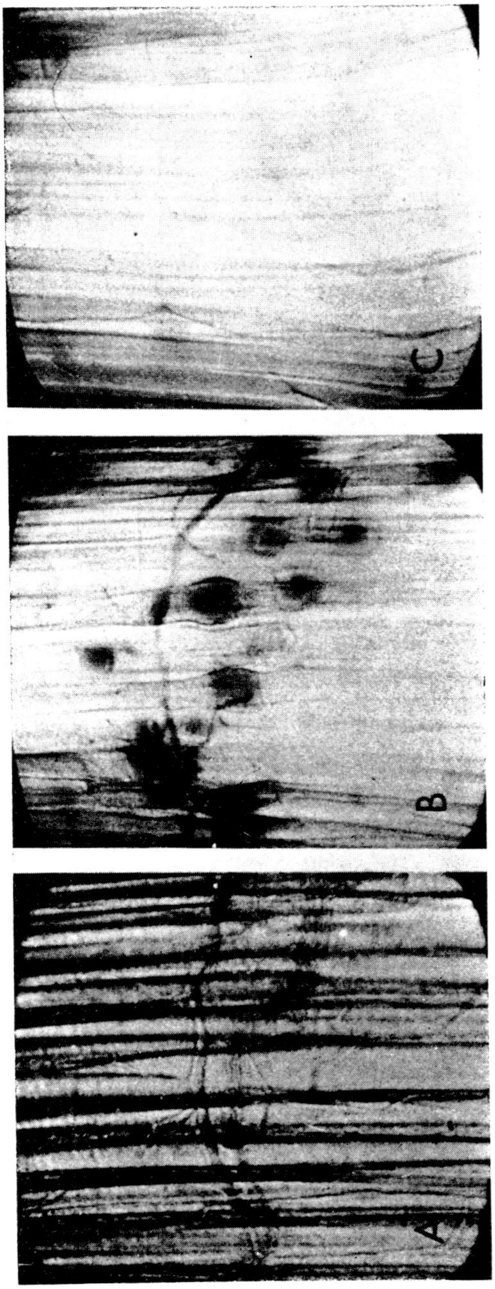

Figure 3–17. *A*, Photomicrograph of a portion of an isolated frog sartorius muscle. A blood vessel runs horizontally across many muscle fibers. Immediately below this blood vessel one can see a bundle of motor neurons which travel across many muscle fibers and forms neuromuscular junctions with some of them. *B*, Same region as in *A* but after exposure of the muscle to choline (5 mM) and then to alizarin red S. The darkened fusiform areas represent junctional regions stained orange-red by precipitated Ca-alizarin red S. *C*, Nonjunctional region of the same muscle treated as described in *B* above. Further details in reference 10.

REFERENCES

1. Nastuk, W.L. (Ed.): *Physical Techniques in Biological Research. Electrophysiological Methods.* New York, Academic Press, 1964, vol. 5.
2. Karis, J.H.; Gissen, A.., and Nastuk, W.L.: Mode of action of diethyl ether in blocking neuromuscular transmission. *Anesthesiology, 27:*42–51, 1966.
3. Nastuk, W.L., and Karis, J.H.: The blocking action of hexafluorenium on neuromuscular transmission and its interaction with succinylcholine. *J Pharmacol Exp Therap, 144:*236–252, 1964.
4. Nastuk, W.L., and Gissen, A.J.: Action of certain quaternary ammonium compounds in neuromuscular transmission. In Paul, W.M.; Daniel, E.E.; Kay, C.M., and Monckton, G. (Eds.): *Muscle Structure and Function.* New York, Pergamon, 1965.
5. Nastuk, W.L.: Activation and inactivation of muscle postjunctional receptors. *Fed Proc 26:*1639–1646, 1967.
6. Manthey, A.: The effect of calcium on the desensitization of membrane receptors at the neuromuscular junction. *J Gen Physiol, 49:*963–976, 1966.
7. Nastuk, W.L., and Parsons, R.L.: Factors in the inactivation of postjunctional membrane receptors of frog skeletal muscle. *J Gen Physiol, 56:*218–249, 1970.
8. Katz, B., and Thesleff, S.: A study of the 'desensitization' produced by acetylcholine at the motor end-plate. *J Physiol, 138:*63–80, 1957.
9. Marco, L.A.: Mikiten, T.M., and Nastuk, W.L.: Histochemical detection of calcium released at neuromuscular junctions of oscillating frog muscle fibers. *Fed Proc, 28:*457, 1969.
10. Mikiten, T.M.; Marco, L.A., and Nastuk, W.L.: Calcium release and sarcomeric oscillations at the neuromuscular junction of frog sartorius muscle. Unpublished results.
11. Nastuk, W.L.: Mechanisms of neuromuscular blockade. *Ann NY Acad Sci, 183:* 171–182, 1971.

Chapter 4

Muscle Relaxants

Stuart C. Cullen

When curare was first used in medical practice, its potential for beneficial effect and its inherent danger to the organism into which it was introduced were well recognized. Until 1939, curare was an enigmatic, awesome, yet intriguing lump of material which inhabited many a laboratory. From the days of Sir Walter Raleigh and Fontana, scientists were fascinated with the tales of explorers who visited the Indians in South America, saw the lethal preparations used by them in hunting, and were impressed with the potency of this concoction. Could one risk investigating a drug so steeped in witchcraft let alone applying it to human beings? Nevertheless, the crude masses of the materials made available to scientists by explorers were investigated. A reliable mode of action was determined and the unpredictable preparation introduced into human beings for the treatment of numerous disorders.

The men who had the imagination to take the information supplied by the scientists in the laboratory and apply it to patients with disorders which seemed, at the time, to be appropriately susceptible to the effects of the drug were such people as West, Berman, Mitchell, Benedict, and Hunter. The disorders which seemed largely vulnerable to curare were those due to neurological lesions associated with trauma, tetanus, strychnine poisoning, cholera, rabies, Parkinson's disease, hemiplegia, hypertension, and epilepsy. The literature reports that the drug in its crude form was applied to patients with these sorts of disorders with varying degrees of success. The success or failure of these courageous ventures into new therapeutics was not so much a function of the appropriateness or inappropriateness of the therapy as it was a function of the unreliability of the preparation and the inability of the therapist to quantitate the dose. In addition, they were unprepared to compensate the respiratory insufficiency associated with the effective dose. They were either unaware of, or skeptical of, the useful investigations of Simpson and Waterton in which they established, by application of curare to horses, that life could be maintained if adequate ventilation were performed.

Brody had determined earlier that the drug affected respiration primarily. This research was outstanding not only in its relation to the safe application of curare and other relaxants, but also in relation to the demonstration of the value of artificial ventilation. A technique of ventilation used by Simpson and Waterton was almost identical to that found to be most effective in light of accurate data available to date on the parameters of artificial ventilation. All of the men who worked with these original investigators had profound respect for the potency as well as the potential of the preparation.

In this same atmosphere of inquisitiveness and caution, Richard Gill managed, with the help of MacIntyre and Bennett, to overcome the reluctance of scientists in the Squibb laboratories to the extent that they prepared a suitable and predictable product from botanical specimens which Gill had brought with him from Equador. It is of interest that the same fundamental explorer-scientist relationship existed in the twentieth century between Gill, MacIntyre, and Bennett as existed in the seventeenth and eighteenth centuries between Raleigh and Fontana.

It is not surprising that Gill had difficulty in trying to promote the development of curare as a useful drug in modern clinical medicine. He, as well as all the others involved in fostering this phenomenal drug, was consistently concerned about the hazards associated with the widespread application of curare. He recognized it as a drug with lethal potential, especially if used in an indiscriminate fashion or by persons unaware or unconcerned with the disruption of normal physiological mechanisms. Apprehension as well as enthusiasm must have plagued these men. Each of them had faith in the potential of curare and that faith has been substantiated after many years of widespread application in many different aspects of clinical medicine. In addition, there has been a significant bonus not fully unanticipated in the improved understanding of neuromuscular transmission made possible by the predictability of a stable product and a tremendous interest generated by the obvious benefit to medicine.

The reputation of curare as a therapeutic agent has been good. Part of this excellent reputation may have been the result of basic principles outlined and emphasized early in the preparation of the drug. These same principles have been reiterated with the presentation of other relaxants which are not anesthetic, analgesic, or sedative drugs. It was recommended that these drugs be used only in those circumstances in which muscle relaxation was the indication. The muscle relaxants were expected at the time of their introduction to clinical anesthesia to be an adjunct to anesthesia and not a substitute for it.

These precepts were as follows:

1. Anesthesia should be established within the limits of safety of the anesthetic drug or combination of drugs.

2. The relaxants should be administered only after the anesthetic had been established at a point of maximum efficiency.
3. Surgical procedures requiring relaxation could be accomplished with agents such as nitrous oxide or ethylene in combination with the relaxants, provided maximum efforts were made to secure optimum anesthesia with any of the less potent agents.
4. Higher and sometimes deleterious concentrations with the more potent anesthetic agents could be avoided by inducing necessary relaxation with a relaxant.
5. Restriction of ventilation was recognized as a usual accompaniment of the use of relaxants.
6. Hypotension, especially in a poor-risk patient to whom large doses were administered, was recognized as a possibility and suggestions for prevention and treatment of this complication were published.

All of these things appeared in early literature associated with the introduction of curare into clinical anesthesia. It was anticipated that abuses would accompany the use of relaxants. These abuses in general are a function of an effort, either deliberate or inadvertent, to use relaxants as substitutes for anesthesia rather than as adjuncts to anesthesia. Concern was expressed that the relaxants would be used to cover up deficiencies in the management of anesthesia. For the anesthetist whose primary object in anesthesia management is to satisfy the acute demands of the surgeon regardless of the impact on the patient, curare and other relaxants are a boon. Such an anesthetist chooses to ignore or fails to recognize the source of the following complications: laryngospasm, exaggerated diaphragmatic action associated with upper respiratory obstruction, and inadequate ventilation. Relaxants were a panacea for these conditions. Anesthetists of this caliber and attitude seem not concerned about the production of a suitable anesthetic state and produce instead a paralyzed rather than anesthetized patient. There have been reports of situations in which a patient was made inactive but not insensible, a situation not always appreciated by the patient although seemingly satisfactory for many anesthetists and surgeons. The inept or indifferent anesthetist who leans inordinately upon the relaxants for a solution to his problems in anesthesia often fails to become aware of the fact that most surgeons made happy at the time of surgery are distinctly less appreciative when the patient succumbs as a consequence of the mismanagement of relaxants.

One of the most subtle abuses is overdosage. Other than hypoventilation, there are a number of immediate deleterious effects from relaxants, and anesthetists often err on the side of too much rather than the side of too little. It is established that, although proper attention to ventilation can obviate one of the hazards of overdose during the period of anesthesia, there is a significant hangover effect in the postanesthetic period.

Prior to the introduction into clinical practice of application of electrical response, most of us relied on the headlift test or squeezing of the hand or any of these sorts of clinical maneuvers to judge the adequacy of muscle strength following a surgical procedure. With the advent of the application of electrical stimulation as proposed by Churchill-Davidson, it has become quite evident that these simple clinical tests are inadequate to establish that there is insufficient relaxant left to cause problems. The electrical stimulation test is superior in any circumstance in which large volumes of relaxants are used if there is any question of leftover or hangover effect of the relaxant. By using this test, the chances of a patient being in serious difficulty because of the occult leftover relaxant effect is significantly reduced.

A residual effect may be expected also with short-acting relaxants. In either case, long or short, the incidence and extent of residual effects is directly proportional to the total dose employed. It is a curious paradox that the anesthetists who should be most concerned with respiratory inadequacy are often the most indifferent to it if it occurs as a consequence of an overdose or leftover effect of the relaxant.

Misadventures in anesthesia are not alleviated by relaxants. On the other hand, they compound the difficulties and although the fatalities attributed to relaxants may have been exaggerated, there is no doubt that they have contributed to mortality. Many anesthetists were distressed at the Beecher-Todd report and felt that it was an exaggeration of the dangers accompanying the use of relaxants. Some surgeons prohibited the use of relaxants for some time because of this Beecher-Todd report. Upon unemotional reflection and analysis of the data, however, it has become abundantly clear that there is a mortality and a morbidity associated with the use of relaxants.

It must be disappointing to those who have been instrumental in introducing these valuable adjuncts to anesthesia to have their previously outlined warnings ignored and to find that relaxants are still used to cover up deficiencies in anesthesia rather than to augment good anesthesia.

In the not too distant past, it was easy to be dogmatic about the anesthetic state and the changes associated with its induction and maintenance. Sleep, analgesia, and relaxation were produced with varying degrees of efficiency by the same agent. Now, as was pointed out by Woodbridge, the definition of an anesthetic state is much more difficult. Coincident with the introduction of curare into clinical anesthesia, was the popularization of the combination of drugs known usually by the term "balanced anesthesia." It was expected that each of the drugs in the mixture contributed to analgesia, sleep, or relaxation resulting in a total anesthetic effect. It became necessary to attempt in each patient to determine whether analgesia, sleep, sedation, or anesthesia was needed in order to supply the appropriate drugs.

Anesthesia is often a conglomerate of various types of depression and the relaxants are an important part of the process of producing suitable working conditions for the surgeon. As different relaxants were made available and as anesthetists became bolder, there was a shift to more liberal use of relaxants without special regard to the presence or absence of the need for muscle relaxation. One of the things that disturbs me more than anything else in supervising anesthesia is to see a patient reacting by grimacing his eyebrows, moving a hand or otherwise indicating light stages of anesthesia and the necessity for more anesthesia, and to see him instead receive more relaxant. In such a circumstance, the anesthetist should increase the concentration of inhalation agent, add more thiopental or narcotic or some other anesthetic agent, but he should not reach for succinylcholine or curare so that the patient will remain paralyzed rather than anesthetized.

The relaxants available today each have specific properties in their ability to produce muscle relaxation. There is information also of a precise nature to facilitate selection of different drugs to fill requirements in different patients. Relaxants can be applied on the basis of known properties rather than on the basis of purely personal preference or habit. One gets the impression that particular relaxants are used exclusively for such quixotic reasons as:

1. It was the first relaxant ever used.
2. It produces sufficient relaxation so why bother with any other.
3. The anesthetist has good luck with it so why should he change.
4. This is the one his favorite drug-detail man recommends.
5. The anesthetist read a paper about it once.
6. The hospital gets the best deal on this particular drug, etc.

I propose also that the state of anesthesia be established prior to and/or concomitant with the relaxant so that the relaxant is used for relaxation and *not* for a substitute of anesthesia. Let us keep in mind that patients need anesthesia and sedation as well as relaxation, and let us continue to exercise care and judgment in the use of the anesthetic agent as well as relaxants.

There may be newer relaxants with precise properties that facilitate management of a particular problem. These new agents may be of such a nature as to produce selective relaxation based on different types of muscles or different susceptibility of the postjunctional membrane. If these are made more available, and this is not unreasonable to expect, the work and the obligations of the anesthetist will be greater if he takes advantage of the increased information for the benefit of the patient. Do not look for a panacea in relaxants but attempt to use, with precision, the relaxants that are available and soon to be introduced.

Chapter 5

Genetic Variability of Enzymes

WERNER KALOW

Current knowledge about the biochemistry of gene action indicates that DNA contains the genetic blueprints for proteins which are then manufactured with the help of RNA and amino acids. Proteins are the primary substances which, besides the genes themselves, may exhibit genetic variability. In other words, variability of anatomical structure, and physiological or biochemical function, including drug response, must be an expression of the variability of enzymes, or perhaps more broadly speaking, of functional proteins. With these premises laid down, I shall first comment on the prevalence of genetically variable enzymes in man. In the second part of the paper, I will review examples of enzyme polymorphism which influence drug metabolism.

The question of rarity or abundance of enzyme variants does comprise a number of problems, such as how many enzymes are variable, how often does one find multiple forms of a variable enzyme, are persons with an enzyme variant necessarily rare in a population, is the incidence the same in different populations. One can answer some of these questions.

The only systematic study designed to give an estimate of the number of genetically variable enzymes in man was conducted by Professor Harry Harris of the Galton Laboratory in London.[6] He investigated eighteen enzymes of human blood by starch gel electrophoresis, the choice of enzymes being determined by a lack of previous genetic studies and by detectability of the enzyme on the gel. Counting only genetic variants which occurred more frequently than in 1 per cent of the investigated population, he found six of the eighteen enzymes genetically variable. To interpret this proportion, one has to keep in mind that not all variants would show up by starch gel electrophoresis. Variants characterized by a change of one amino acid for another would be detectable by electrophoresis if the amino acids have different charges, but not if they have equal charges. Thus if Harris' data are representative, they imply that at least half of all human enzymes are genetically variable. Table 5-I is the reproduction of Harris' list of enzymes.

Table 5-II contains a list of enzymes of human red cells which have been shown to be genetically variable.[1] It is a substantial list. The many other genes that the prevalence of a variant is not alike in different populations. For instance, most specific variants of glucose-6-phosphate dehydrogenase are confined to certain areas of the world. Among persons of Northern European descent, any variant is rare; while in some parts of the near East, 30 to 40 per cent of the population is affected.

If an enzyme variant occurs in different populations with unlike but appreciable frequencies, one may assume that the variation is functionally not neutral, and that the variant conveys in one surrounding a greater selective advantage than in another. Perhaps one should not so much stress the selective advantage, as the balance between advantage and disadvantage. A heterozygote is often favored in a surrounding in which the homozygote has a disadvantage. The famous example in human genetics is the explanation for the abundance of hemoglobin S, the hemoglobin variant of sickle cell anemia, in some populations of tropical origin. Heterozygotes with one sickle cell gene withstand malaria better than do persons with only the standard hemoglobin, while the homozygous person with the double dose of hemoglobin S is likely to die from his anemia. In most other cases in man, the nature of the advantage or disadvantage is not known. Very rare genes are likely rare because they are freshly produced from time to time by new mutation but disappear quickly because they convey a disadvantage in most surroundings.

One may ask how many individuals can be regarded as normal in the sense that they have the most common phenotypes of variable enzymes. Harris' estimate[6] is shown in Table 5-IV. It turns out that less than 2 per cent of the individuals of the investigated population can be regarded as normal in respect to the listed enzymes. The probability of two randomly selected individuals being of the same type is less than 0.5 per cent. Consider that there may be innumerable undefined enzymes in the human body. Thousands of enzymes have been described, but only a small fraction of these have been tested for genetic variability. Of these, only the few that show a common polymorphism are taken into consideration. On the basis of these few, almost every person is enzymologically different from the next.

There are many relations between enzymes and drug action. Drugs may be subject to biotransformation by enzymes. The metabolic alteration of the drug may mean inactivation and thereby termination of action, or it may mean conversion of an inactive to an active chemical. Furthermore, drugs may interact with enzymes by being substrates, inhibitors, or activators of enzymes. Only in recent years has one realized that drugs may affect rate of formation of enzymes and thereby in-

TABLE 5-IV
ENZYME POLYMORPHISM IN THE ENGLISH POPULATION.
(DATA ON TEN LCOI)

Enzymes	*Number of Alleles with Frequency Greater than 0.01*	*Frequency of Commonest Phenotype*	*Probability of Two Randomly Selected Individuals' Being of Same Type*	*Proportion of Population Who Are Heterozygous*
Red-cell acid phosphatase	3	0.43	0.34	0.51
Phosphoglucomutase				
Locus PGM_1	2	0.58	0.47	0.36
Locus PGM_2	2	0.55	0.45	0.38
Placental alkaline phosphatase	3	0.41	0.30	0.50
Liver acetyltransferase	2	0.50	0.50	0.50
Adenylate kinase	2	0.90	0.82	0.10
Serum cholinesterase				
Locus E_1	2	0.96	0.92	0.04
Locus E_2	2	0.90	0.82	0.10
Phosphogluconate dehydrogenase	2	0.96	0.92	0.04
Adenosine deaminase	2	0.88	0.79	0.11
All enzymes combined	—	0.018	0.005	—

(Courtesy of Professor Harry Harris and the *British Medical Bulletin*)

fluence enzyme levels. The facts of interaction between drugs and enzymes, and the genetic variability of enzymes lead to the conclusion that there should be much genetic variability of drug action. This is indeed observed and often referred to as pharmacogenetics. While this topic has many facets, limitations of time will allow me to consider only one, namely drug inactivation by genetically variable enzymes.

One pertinent example is the occurrence of genetically variable forms of cholinesterase in human plasma. The unusual variants fail to effectively hydrolyse succinylcholine so that it persists and exerts a prolonged action. It has been known for many years that there are a number of different variants; this fact indicates that detailed genetic typing of a person in regard to cholinesterase is not necessarily an easy chore. However, it has become easy to determine whether or not there is an enzyme abnormality. Motulsky and collaborators[8] have recently published a method of detection of cholinesterase abnormalities which seems to constitute an optimal compromise between ease of determination and resolving power of the test. The question has been raised repeatedly whether such a test should be used to screen patients who are to receive succinylcholine. I am not convinced that much would be gained by such a routine pretesting. Under most circumstances, artificial respiration will tide the patient over some period of apnea without much risk. Only if a large dose of succinylcholine is needed, should one use either the biochemical test for cholinesterase or inject a small test dose prior to infusion of a large amount. However, relatives of affected patients should be tested biochemically; one is so likely to encounter esterase variants among them that an esterase test is worthwhile. Goedde[5] has made therapeutic trials to terminate prolonged apnea by injection of a large amount of purified and concentrated human cholinesterase. One can shorten a prolonged apnea by this process. However, I am not presently aware of anyone having prepared pure cholinesterase. One would have to weigh the dangers of an allergic reaction to injected protein against the inconvenience of having to provide artificial respiration for a limited period.

In the field of drug metabolism, the interaction between succinylcholine and cholinesterase is unusual, if not unique. A great majority of drug-metabolizing enzymes are located intracellularly, mostly in liver cells in structures which in purified state are referred to as microsomes. As a rule, one must assume that ionized drugs like succinylcholine have a limited access to intracellular enzymes. Yet, the same ionization makes succinylcholine a particularly suitable substrate for the extracellular cholinesterase. The fact that succinylcholine is usually given intravenously helps it to be more quickly destroyed in plasma than other potential substrates of cholinesterase which are administered by a different route.

In view of the unusual nature of succinylcholine inactivation, it deserves emphasis that some of the common drug-inactivating enzymes in human liver have been shown to be under hereditary control. A few years ago, Vesell advertised in the *Washington Post* for twins to volunteer for drug studies. There was enough response to permit him to be selective. He ended up with a number of fraternal and identical twins, all of whom were Caucasian adults and most of whom were living in separate households apart from each other. This lack of common environment made the study particularly valuable. The drugs which Vesell and his collaborators[10] gave were phenylbutazone, antipyrin, and dicoumarol. They studied the rate of disappearance from plasma. It turned out that the half-life of each drug was similar within pairs of identical twins and variable within pairs of fraternal twins (Table 5-V). The data indicate that the persistence of all three drugs in the human body was under genetic control. This result came as a surprise, since in animal experiments the metabolism of all three drugs is known to depend on adaptable enzymes in liver microsomes.

The twins in Vesell's study were instructed prior to the experiment not to take any drugs, to avoid enzyme inducers like barbiturates. In the meantime, it has been described that nicotine is able to enhance the activity of drug-metabolizing enzymes, and the same claim has been made for caffeine. Vesell's data clearly indicate that nicotine and caffeine intake had no major influence on the metabolism of the three drugs.

TABLE 5–V
DICOUMAROL, ANTIPYRINE, AND PHENYLBUTAZONE HALF-LIVES WITH SMOKING AND COFFEE HISTORY IN TWENTY-EIGHT TWINS

Twin	*Age, Sex*	*Half-life* *Dicoumarol*	*Antipyrene*	*Phenylbutazone*	*Smoking*	*Coffee*
		hr	hr	days	pack / day	cups / day
			IDENTICAL TWINS			
Ho.M.	48, M	25.0	11.3	1.9	0.5	2
Ho.M.	48, M	25.0	11.3	2.1	1	3
D.T.	43, F	55.5	10.3	2.8	0	5–6
V.W.	43, F	55.5	9.6	2.9	2	8-10
J.G.	22, M	36.0	11.5	2.8	1	1–2
P.G.	22, M	34.0	11.5	2.8	1	1–2
Ja.T.	44, M	74.0	14.9	4.0	0	6
Ja.T.	44, M	72.0	14.9	4.0	0	2–3
C.J.	55, F	41.0	6.9	3.2	0	2
F.J.	55, F	42.5	7.1	2.9	0	2
Ge.L.	45, M	72.0	12.3	3.9	0	4
Gu.L.	45, M	69.0	12.8	4.1	0	4
D.H.	26, F	46.0	11.0	2.6	0	0–1
D.W.	26, F	44.0	11.0	2.6	0	3–4
			FRATERNAL TWINS			
A.M.	21, F	45.0	15.1	7.3	1.5	2
S.M.	21, M	22.0	6.3	3.6	0	0
D.L.	36, F	46.5	7.2	2.3	0	2–3
D.S.	36, F	51.0	15.0	3.3	2	3–4
S.A.	33, F	34.5	5.1	2.1	1	2
F.M.	33, F	27.5	12.5	1.2	0.5	2
Ja.H.	24, F	7.0	12.0	2.6	0	10–15
Je.H.	24, F	19.0	6.0	2.3	1.5	10
F.D.	48, M	24.5	14.7	2.8	0	1
P.D.	48, M	38.0	9.3	3.5	1.5	8
L.D.	21, F	67.0	8.2	2.9	1	6
L.W.	21, F	72.0	6.9	3.0	1	2–3
E.K.	31, F	40.5	7.7	1.9	0	0
R.K.	31, M	35.0	7.3	2.1	1	0

(Courtesy of Dr. Elliot S. Vessell and the *Journal of Clinical Investigation*)

There is, however, one factor that forces one to be cautious in interpreting the data in terms of drug metabolism. It has been assumed that all drugs would be metabolized by hydroxylation. There was a distinct correlation between half-lives of phenylbutazone and dicoumarol but none between antipyrine and any of the other drugs. It is possible, therefore, that the half-lives of dicourmarol, and phenylbutazone were determined not so much by their rate of metabolism but by the rate of release from plasma protein binding. Both dicourmarol and phenylbutazone are known to be extensively bound to plasma albumin. Yet this potential objection should not pertain to antipyrine, so that there is no reason to doubt the principal finding of genetic control of microsomal enzyme activity under ordinary circumstances.

There are older data indicating that microsomal metabolism of drugs may be genetically controlled in rare families. In 1964, Kutt and co-workers[7] found a family in which rate of metabolism of diphenylhydantoin was clearly familiar. The central effects of this drug were related to genetically defective metabolism. Diphenylhydantoin is hydroxylated in man. More recently another family study has been reported by Shahidi which concerns acetophenetidin or phenacetin.[9] Shahidi's discovery reads like a detective story. He investigated a young woman who, suffering from methemoglobinemia, repeatedly entered a hospital. She denied having ever taken drugs, but it was finally discovered that she took in large doses suppositories containing phenacetin. It then turned out that this woman metabolized phenacetin by ring-hydroxylation, which converted the drug into a methemoglobin-forming compound. Usually phenacetin is metabolized by one or two different processes of cleavage. In short, the drug metabolism proceeded along an unusual pathway which led to the formation of a poisonous substance. It then turned out that a sister of this woman showed the same unusual pattern of phenacetin metabolism, thus suggesting recessive inheritance.

I described earlier that the ability to acetylate isoniazid divides the North American population into almost equal groups of rapid and slow acetylators. This polymorphism is determined by the activity of an acetyltransferase in human liver. Besides isoniazid, this enzyme also acetylates the mono-amino-oxydase inhibitor Nardil® and the sulfonamide sulfadimidine. It is interesting to note that sulfanilamide itself is acetylated by a less variable and different acetyltransferase, presumably in the intestinal wall. Acetylation is not a pathway of destruction for very many drugs.

Conjugation of drugs with glucuronide depends on glucuronyl transferases, which have often been discussed because a lack of glucuronidation in infants is a cause of kernicterus. The enhancement of kernicterus by drugs has not much to do with the lack of glucuronidation of the drugs themselves but with overloading of the plasma albumin which diminishes the carrying capacity for unconjugated bilirubin. Also, the intoxication of infants with chloramphenicol, descriptively called the "gray syndrome," is due to a lack of glucuronidation in combination with underdeveloped renal function. Many drugs are metabolized by glucuronidation but quite often glucuronidation is secondary to some other metabolic transformation of the drug and thus is rarely directly connected with termination of pharmacological action.

Von Wartburg and his collaborators[11] found some years ago an unusual form of alcohol dehydrogenase in human liver. Since genetic studies on human liver enzymes are hard to justify the hereditary nature of this enzyme polymorphism has not been proven but can hardly have any other but a genetic cause. The unusual alcohol dehydrogenase is capable of metabolizing alcohol five times as fast as is the normal enzyme. It was then observed that this biochemical difference did not make much difference for the alcohol metabolism in the living subject. This paradox seems to be explained by the fact that the alcohol dehydrogenase activity is limited by the reoxidation of the nucleotide $NADH_2$ rather than by the enzyme itself.

Concluding this chapter, one may say that there is abundant evidence for a genetic variation of many drug-metabolizing enzymes in man. The clinical significance of this variation differs for each enzyme and for each drug. It is particularly interesting to see that there is genetic control of human liver enzymes which are presumably also subject to adaptive enzyme formation. The relative importance of such adaptation and hereditary predisposition is thus still to be elucidated.

There are numerous variants of enzymes other than those concerned with drug metabolism, which are important to pharmacology. One of the most prominent examples is the occurrence of hemolysis due to deficiency of glucose-6-phosphate dehydrogenase. I have mentioned before that there are a great number of different variants of this enzyme, and that deficiency of several enzymes other than glucose-6-phosphate dehydrogenase may lead to hemolysis. One must look at glucose-6-phosphate dehydrogenase deficiency as one example of an enzymatic aberration which sensitizes a person to have a toxic reaction to an otherwise innocuous drug. There are many other examples of an increased susceptibility to toxic drug reaction on the basis of hereditary predisposition.

In summary, there are numerous examples of genetic alterations of enzymes in man. Perhaps most enzymes are variable, and it could be that none of us is normal in the sense that he has the standard form of all enzymes. Many of the enzyme variations affect drug responses either by causing special vulnerability to some toxic action, or by affecting drug metabolism. It is still most important to be aware that dangerous drug reactions can happen on a genetic basis. For persons who wish to hunt for hereditary enzyme defects there must be still large game to be bagged.

REFERENCES

1. Beutler, E. (Ed.): *Hereditary Disorders of Erythrocyte Metabolism*. New York, Grune and Stratton, 1968.
2. Beutler, E.: Drug-induced hemolytic anemia. *Pharmacol Rev, 21*:73, 1969.
3. Carson, P.E.: Hemolysis due to inherited erythrocyte enzyme deficiences. *Ann NY Acad Sci, 151* (Art 2):765, 1968.
4. Evans, D.A.P.: Pharmacogenetics. *Am J Med, 34*:639, 1963.
5. Goedde, H.W.; Altland, K., and Scholler, K.L.: Pharmakogenetische Reaktion auf

Succinyldicholin: Therapie der Verlängerten Apnoe. *Med Klin, 62:*1631–1635, 1967.

6. Harris, H.: Enzyme and protein polymorphism in human populations. *Br Med Bull, 25:*5, 1969.
7. Kutt, H.; Wolk, M.; Scherman, R., and McDowell, F.: Insufficient parahydroxylation as a cause of diphenylhydantoin toxicity. *Neurology* (*Minneap*), *14:*542, 1964.
8. Morrow, A.C., and Motulsky, A.G.: Rapid screening method for the common atypical pseudocholinesterase variant. *J Lab Clin Med, 71:*350, 1968.
9. Shahidi, N.T.: Acetophenetidin-induced methemoglobinemia. *Ann NY Acad Sci, 151* (Art. 2):822, 1968.
10. Vesell, E.S., and Page, J.G.: Genetic control of Dicumarol levels in man. *J Clin Invest, 47:*2657, 1968.
11. Von Wartburg, J.P., and Schürch, P.M.: Atypical human liver alcohol dehydrogenase. *Ann NY Acad Sci, 151* (Art. 2):936, 1968.

Comprehensive information on pharmacogenetics has been compiled recently with B.N. LaDu and W. Kalow as scientific editors in *Ann NY Acad Sci, 151*(Art. 2), 1968.

Chapter 6

The Detection of Genetic Carriers in Hereditary Metabolic Disorders

David Yi-Yung Hsia

One of the exciting new developments in human genetics has been identification of genetic carriers for abnormal biochemical traits. If an individual is heterozygous for a given trait, he carries one "normal" and one "mutant" gene and the molecules derived from the "mutant" gene may be expected to differ from those derived from the "normal" gene within the same individual. On superficial examination, such individuals might appear to be asymptomatic. However, more careful physical and biochemical studies may reveal departures from the norm and these individuals can frequently be distinguished from the population as a whole.

As a result, the concepts of dominance and recessiveness can be regarded largely as operational definitions, which for medical and counseling purposes can be quite useful. However, genetically speaking, most conditions are not transmitted as completely dominant or recessive, but relatively so, depending upon the ease with which the character can be detected among heterozygotes. Thus, phenylketonuria may be classified as an incompletely recessive condition; sicklemia, as being neither dominant nor recessive, but intermediate; and hereditary spherocytosis, as incompletely dominant.

We shall review several of the biochemical methods which have been used in the detection of heterozygous carriers in hereditary metabolic diseases.

1. Direct Measurement of Enzymes

Galactosemia is a hereditary metabolic disease characterized by an inability to convert galactose to glucose in a normal manner. Infants with this disorder appear to be normal at birth; but after a few days of milk feeding, they begin to vomit, become lethargic, fail to gain weight, and show enlargement of the liver and jaundice. Biochemical studies have shown that there is a deficiency of galactose-1-phosphate uridyl transferase in the red blood cells, white blood cells, and other

These studies were aided by grants from the Illinois Mental Health Fund and the United States Public Health Service (HD 04346)

tissues in such patients. In Figure 6–1, we have listed the actual values for galactose-1-phosphate uridyl transferase using the UDPG consumption test in thirty-nine patients with galactosemia (homozygous affected), fiftyfour parents of galactosemic children (heterozygotes), and forty-four normal controls. It can be seen that there is a minimum of overlap between the three groups, with galactose-1-phosphate uridyl transferase for the heterozygotes ($12.0\mu M \pm 3.3\mu M$ UDPG/hr/gm Hgb) being almost

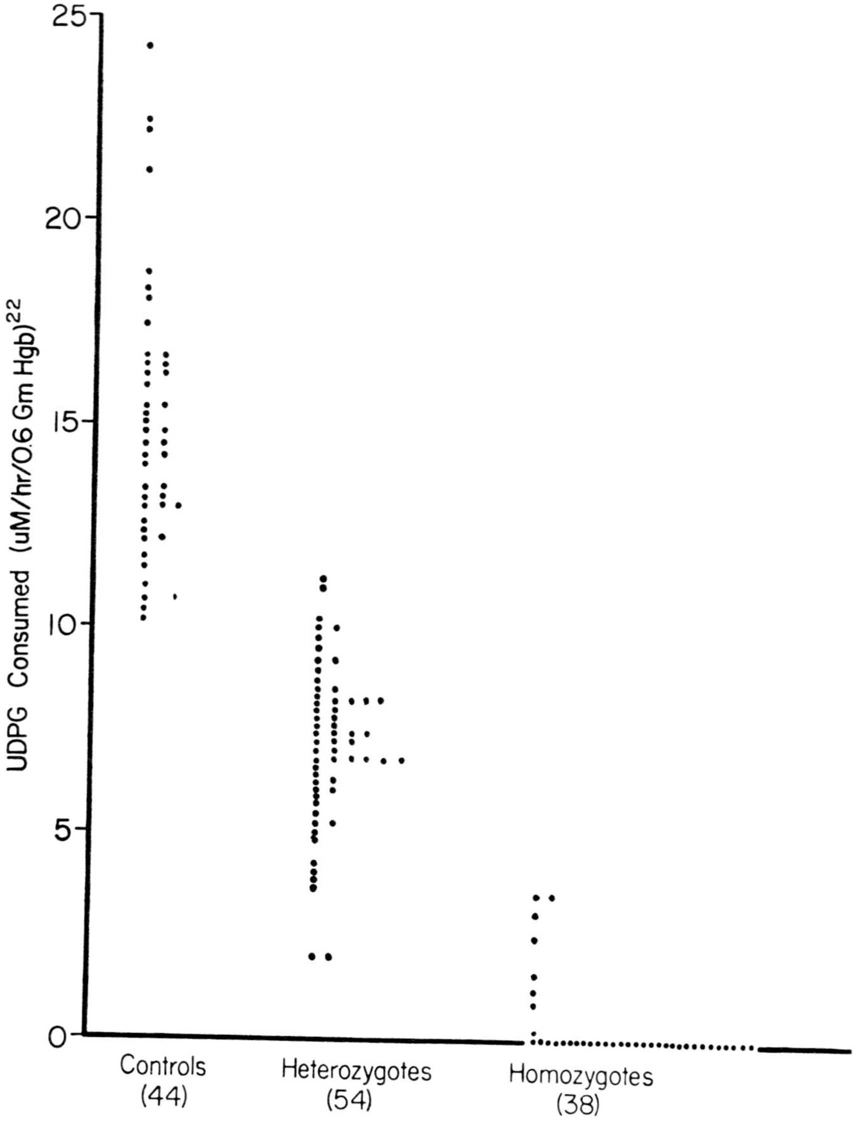

Figure 6–1. Erythrocyte galactose-1-phosphate uridyl transferase activity in controls, heterozygotes, and homozygotes in thirty-seven galactosemic families.

exactly half that for the controls (25.0μM $\pm$ 5.3μM UDPG/hr/gm Hgb).

The direct measurement of enzyme or metabolite in heterozygous carriers has been successfully achieved in a number of "inborn errors of metabolism" using red blood cells, white blood cells, plasma or serum, urine, and tissues. It has also been possible to detect the heterozygous female carriers of X-linked traits using different cell and tissue systems.

2. *Loading Tests*

Unfortunately, it is not always possible to demonstrate a difference between heterozygous carriers and normal controls in an autosomal recessive trait by using such simple and elegant methods. For example, in phenylketonuria, the homozygous affected individual has been shown to have a deficiency of phenylalanine hydroxylase in the liver. The heterozygous carrier for the abnormal gene would be expected to have about half of the normal enzyme activity, but this cannot be easily detected by simple clinical or biochemical means. Such individuals are not necessarily blond or blue-eyed and seldom are mentally retarded. When examined at random, their plasma phenylalanine is only minimally increased above control values and their urine shows the presence of phenylpyruvic acid and o-hydroxyphenylacetic acid only intermittently. Since it would be difficult to get liver biopsies performed on healthy parents of phenylketonuric children, we have no data on phenylalanine hydroxylase activity among heterozygotes for phenylketonuria.

However, when such parents are given a load of 0.1 gm/kg body weight of L-phenylalanine by mouth, they are not able to handle this amino acid as satisfactorily as a normal person. As shown in Figure 6–2, the heterozygotes have plasma phenylalanine levels on the average twice those seen among controls, suggesting a defect in the hydroxylation of phenylalanine.

Loading tests have also been found to be useful in the detection of heterozygous carriers in a number of inborn errors of metabolism.

3. *Cell Cultures*

During the last few years, increased attention has been turned to the use of cell culture techniques for the identification of heterozygous carriers. These developments may be considered under several headings: (1) direct measurement of enzymes and metabolites in fibroblasts, (2) identification of carriers by the presence of metachromatic granules, (3) investigation of cultivated fetal cells derived from amniotic fluid, and (4) cloning of fibroblasts, grown from female heterozygous carriers of X-linked traits.

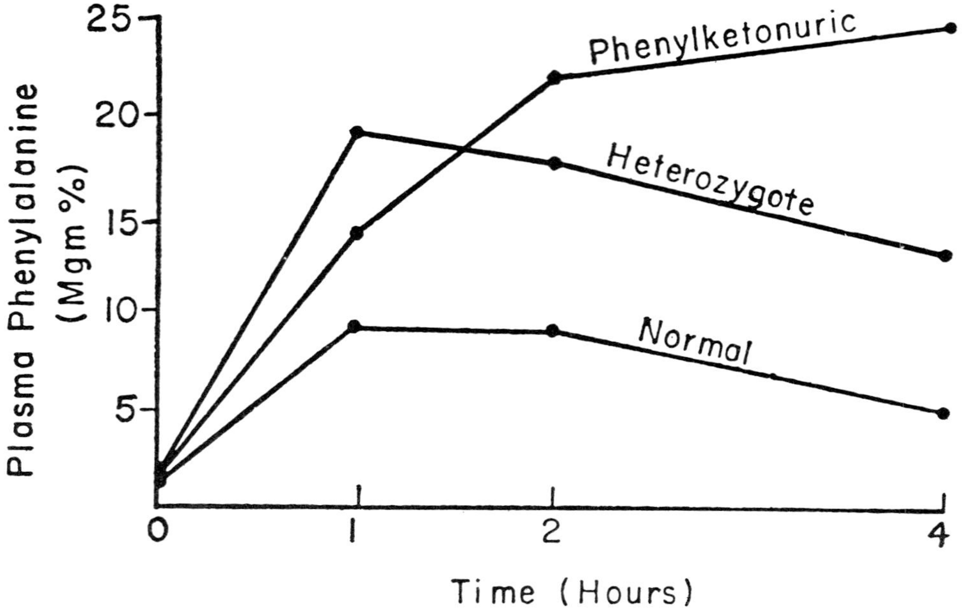

Figure 6–2. Phenylalanine tolerance tests on parents (heterozygotes) of phenylketonurics as compared with normal controls. The curve of a phenylketonuric is also shown for comparison.

For the studies with fibroblasts, skin obtained by biopsy is cut into several pieces and immobilized under cover slips in 35 mm Petri dishes containing culture medium (nutrient mixture F10, Grand Island Biological Co.; 15% fetal calf serum, 50μg/ml of penicillin, and 50μg/ml of streptomycin) and placed in a carbon dioxide incubator. Replicate aliquots of the appropriate cell suspensions are introduced into plastic Petri dishes and sampled at various intervals during growth for determination of enzyme activity and protein content.

In Table 6-I, we have tabulated the activity of galactose-1-phosphate uridyl transferase in fibroblasts from normals, heterozygotes, and homozygotes for galactosemia based upon the data of Krooth. It can be seen that the ratio of Gal-$^{14}C \rightarrow {}^{14}CO_2$ to Glu-$^{14}C \rightarrow {}^{14}CO_2$ for heterozygotes lies between that of normals and galactosemics, much in the same manner as by direct measurement of red blood cells and white blood cells. It is of some interest and importance to note that the metabolic changes in skin fibroblasts tend to parallel closely those seen with white blood cells.

Not infrequently, leukocytes and fibroblasts provide a better measure of tissue enzyme activity and demonstrate the biochemical defect which is not present in the erythrocytes. For example, in Type II glycogen storage disease (Pompe's disease), the enzyme defect showing

TABLE 6–I
ACTIVITY OF GALACTOSE-1-PHOSPHATE URIDYL TRANSFERASE IN FIBROBLASTS FROM NORMALS, HETEROZYGOTES, AND HOMOZYGOTES FOR GALACTOSEMIA

	(1) Gal-^{14}C → $^{14}CO_2$	(2) Glu-^{14}C → $^{14}CO_2$	Ratio (1) / (2)
Normal strains (G^+/G^+)	352	807	0.4
	87	298	0.3
	157	527	0.3
	1235	3544	0.3
	266	944	0.3
Heterozygotes (G^+/G^G)	67	720	0.09
	66	735	0.09
Homozygotes (G^G/G^G)	0	—	—
	7	590	0.01
	0	2644	0
	1	4541	0

Based on the data of Krooth.

a deficiency of L-glucosidase was first described in liver and muscle, and no abnormality could be found in the erythrocytes. Subsequent studies have shown that there is a deficiency of L-glucosidase in both the leukocytes and fibroblasts of homozygous patients, and that parents of such infants show about half of the normal amount of enzyme.

Recently, cultured fibroblasts have also been found to be useful for the identification of metachromasia among heterozygous carriers for a number of disorders associated with mucopolysaccharide metabolism. For example, in cystic fibrosis of the pancreas, it has been shown that although the homozygous-affected patients have a marked elevation of sweat chloride, the parents of such childern have only a minimal increase of sweat chloride over normal adults. However, if fibroblasts derived from such individuals are stained, both homozygous-affected and heterozygous carriers for cystic fibrosis will show metachromasia, while controls will fail to show metachromasia. Present techniques do not permit clear distinction between homozygous-affected and heterozygotes, but they do distinguish them from normals. This technique has now been used for the detection of heterozygous carriers in a number of inborn errors of metabolism, including Gaucher's disease, Batten-Spielmeyer-Vogt syndrome, and Hurler's syndrome and some of its variants.

For studies with fetal cells, 5 to 10 ml of amniotic fluid is collected by amniocentesis using a transabdominal approach. The fluid is centrifuged at 600 g for twelve minutes and the cell pellet resuspended in 0.5 ml of fetal calf serum. One fourth ml of cell suspension is placed in two 30 mm Falcon Petri dishes and a cover slip utilized to immobilize the cells. After twenty or thirty minutes, 2 ml of F10 medium containing 30% calf serum (GIBCO) is placed in the dish and the cells are placed in a carbon dioxide incubator and fed daily. After approximately eighteen days, the cells are trypsinized and subcultured, using extremely dilute trypsin-viokase solutions. The cells are subcultured a second time after three to five days and can be used for biochemical analysis. Theoretically, any condition which can be diagnosed using adult fibroblasts could also be found using cultivated fetal fibroblasts. Several conditions such as galactosemia, glucose-6-phosphate dehydrogenase deficiency, and Pompe's disease have been diagnosed using these techniques before the birth of the infant.

One of the most exciting new developments has been with cloning fibroblasts derived from female heterozygous carriers of X-linked traits. In glucose-6-phosphate dehydrogenase deficiency, the female heterozygous carriers with one "normal" gene and one "mutant" gene for glucose-6-phosphate dehydrogenase will frequently show intermediate enzyme levels. This was shown first with erythrocytes and subsequently with leukocytes and with fibroblasts. In 1963, Davidson, Nitowsky, and Childs found that the cells derived from fibroblasts of such a person could be cloned and subcultured into two populations. In one population, all the cells show full G-6-PD activity, and in the other population, none of the cells showed G-6-PD activity. This not only provided confirmation of the Lyon hypothesis (that one of the X chromosomes is inactive after early cell division), but also shows a fundamental difference in the heterozygous carrier of an X-linked trait. In the former, we are dealing with a 50 per cent enzyme level within two cell populations, one of which has 100 per cent normal enzyme activity and the other has no enzyme activity.

More recently, two cell populations have been described for a number of X-linked traits, including the Hunter form of the Hunter-Hurler syndrome. In this situation, one can use this technique to distinguish with certainty whether in any given family the mucopolysaccharide defect is transmitted in the Hunter form, which is X-linked recessive, or in the Hurler form, which is autosomal recessive. As shown in Table 6-II, the cells from the female heterozygote can be cloned into two population groups in the Hunter form, while this cannot be done in the Hurler form.

REFERENCES

1. Danes, B.S., and Bearn, A.G.: Hurler's syndrome: A genetic study of clones in cell culture with particular reference to the Lyon hypothesis. *J Exp Med, 126:* 509, 1967.
2. Davidson, R.G.; Nitowsky, H.M., and Childs, B.: Determinations of the two populations of cells in the human female heterozygous for G6PD variants. *Proc Natl Acad Sci USA, 50:*481, 1963.
3. Hsai, D.Y.Y.; Driscoll, K.W.; Troll, W., and Knox, W.E.: Detection by phenylalanine tolerance tests of heterozygous carriers for phenylketonuria. *Nature (Lond), 178:*1239, 1956.

TABLE 6–II
RESULTS OF CLONING IN TWO FAMILIES WITH HUNTER'S SYNDROME (X-LINKED RECESSIVE) AND ONE FAMILY WITH HURLER'S SYNDROME (AUTOSOMAL RECESSIVE)

	Original % Meta-chromasia	*Clone No.*	*Clone % Meta-chromasia*
Hunter's (X-linked)			
I. Hemizygous mutant male	94	E	100
Heterozygous female	49	H	100
		A	0
		BB	0
Hemizygous normal (father)	0	N	0
II. Hemizygous mutant male	100	DD	100
Heterozygous female (mother)	45	F	100
		TT	0
		WW	0
Hemizygous normal (father)	0	C	0
Hurler's (Autosomal)			
I. Homozygous-affected	94	G	97
Heterozygous (mother)	34	L	58

Based on the data of Danes and Bearn.

4. Hsia, D.Y.Y.; Huang, I., and Driscoll, S.G.: The heterozygous carrier in galactosemia. *Nature (Lond)*, *182*:1389, 1958.
5. Hsia, D.Y.Y.: *Inborn Errors of Metabolism. Part I. Clinical Aspects,* 2nd ed. Chicago, Year Book Pub, 1966.
6. Hsia, D.Y.Y.: The detection of heterozygous carriers. *Med Clin North Am, 53:* 857, 1969.
7. Krooth, R., and Weinberg, A.M.: Studies on cell lines developed from the tissues. of patients with galactosemia. *J Exp Med, 113:*1155, 1961.
8. Nadler, H.L., and Hsia, D.Y.Y.: Utilization of white blood cells for study of clinical genetic disorders. *Clin Biochem, 1:*192, 1968.
9. Nadler, H.L.: Prenatal detection of genetic defects. *J Pediatr, 74:*132, 1969.
10. Neel, J.V.: The clinical detection of the genetic carriers of inherited disease. *Medicine (Baltimore), 26:*115, 1947.
11. Stanbury, J.B.; Wyngaarden, J.B., and Fredrickson, D.S.: *The Metabolic Basis of Inherited Disease,* 2nd ed. New York, McGraw-Hill, 1966.

Chapter 7

Some Factors Involved in the Use of Certain Pharmacological Agents in the Treatment of Myasthenia Gravis

WILLIAM L. NASTUK, AARON J. GISSEN, AND JOANNES H. KARIS

In a normal individual, neuromuscular transmission is not prone to failure, because it has a large safety margin. Characteristically the patient with myasthenia gravis has little or no transmission reserve and therefore, during repetitive activity, early failure occurs. Such patients suffer from muscular weakness and are impeded from carrying out ordinary voluntary activities.

According to modern concepts, during neuromuscular transmission, acetylcholine (ACh) is released from the motor nerve terminals in packets (quanta), each of which contains thousands of ACh ions. (See Katz[1] for an excellent review.) Several years ago, Elmqvist *et al.*[2] obtained muscle biopsies from patients with myasthenia gravis and they studied various aspects of neuromuscular transmission in single fibers of these tissue preparations. They concluded that in myasthenic muscle, each motor nerve impulse causes release of a normal *number* of ACh quanta. However, each quantum released produced an abnormally small depolarization of the muscle postjunctional membrane (PJM). These local depolarizations, called miniature endplate potentials (MEPPs), were about 0.2 mV in amplitude, which is one-fifth the normal value. Elmqvist *et al.* reasoned that such small sized MEPPs would result if the chemosensitivity of the PJM was abnormally low, but their experimental tests failed to reveal any such insensitivity. They thus decided that the important defect in the myasthenic patient is that each quantum contains an abnormally small amount of ACh, i.e. the quantum size is small. This conclusion is still open to some question in part because it is not easy to determine PJM chemosensitivity in a completely satisfactory manner. Additional experimental work is needed to clear up this and additional related questions.

This work was supported by U. S. Public Health Grants RO1–NS 04988–05 and GM 090–69–07.
We wish to thank Mr. J.T. Alexander for his invaluable technical asistance.
The diethyl neostigmine was supplied by Hoffman-La Roche Inc., Nutley, New Jersey, through the courtesy of Dr. L.O. Randall, to whom we wish to express our appreciation.

Beyond the understanding of the transmission defect in myasthenic patients lies the even more important matter of the mechanism by which such defects are produced. In 1960, Nastuk *et al.*[3] and Simpson[4] proposed that myasthenia gravis is an autoimmune disease, and this idea is widely entertained today. A brief review of the various aspects of autoimmune theory and the status of evidence supporting it was given by Nastuk and Plescia,[5] at the close of the 1965 international symposium on myasthenia gravis.[5] More recently, several investigators have been attempting to produce myasthenia gravis in experimental animals using immunological techniques. It is not possible to give details here; and although some encouraging results have been obtained, the problem is by no means solved.[17]

The clinical management of the myasthenic patient is not a subject we wish to discuss in detail, but others have written at length on this matter. As is well known, the faltering neuromuscular transmission in myasthenic patients can be facilitated by administration of anticholinesterase drugs such as neostigmine or pyridostigmine. The patient must take such drugs continuously and he is often given permission to adjust his dosage within certain limits. On some occasions, a patient may overdose himself with the result that the neuromuscular facilitation is replaced by muscular weakness. This condition, known clinically as cholinergic crisis, is commonly believed to result from accumulation of ACh at neuromuscular junctions whose cholinesterase activity has been deeply inhibited. Another possible mechanism contributing to cholinergic crisis produced by neostigmine is that this drug itself directly depolarizes and then desensitizes the PJM. No *direct* evidence in support of such an action has appeared in the literature. We therefore utilized single-fiber electrophysiological techniques and adjunct methods to determine the direct actions of neostigmine and a diethyl analogue (3-dimethylcarbamoxyphenyldiethyl methylammonium bromide). The rationale for testing the diethyl analogue of neostigmine is that this compound is more active than neostigmine as an anticholinesterase.[6] In addition, work of our own laboratory[7] has shown that with increasing ethyl substitution, analogues of 3-OH phenyltrimethyl ammonium (including edrophonium) showed better capacity to increase and maintain the tension output of indirectly stimulated frog sartorius muscle. Figure 7–1 A-D depicts the results obtained in this study. As the concentration of each compound was raised, the tension output increased (open circles) but with more elevated concentrations, the increase in tension output diminished and was less well maintained. Neuromuscular block appeared with the highest concentration of the trimethyl compound. This compound has the greatest capacity to depolarize and desensitize the PJM. Thus the diethyl ana-

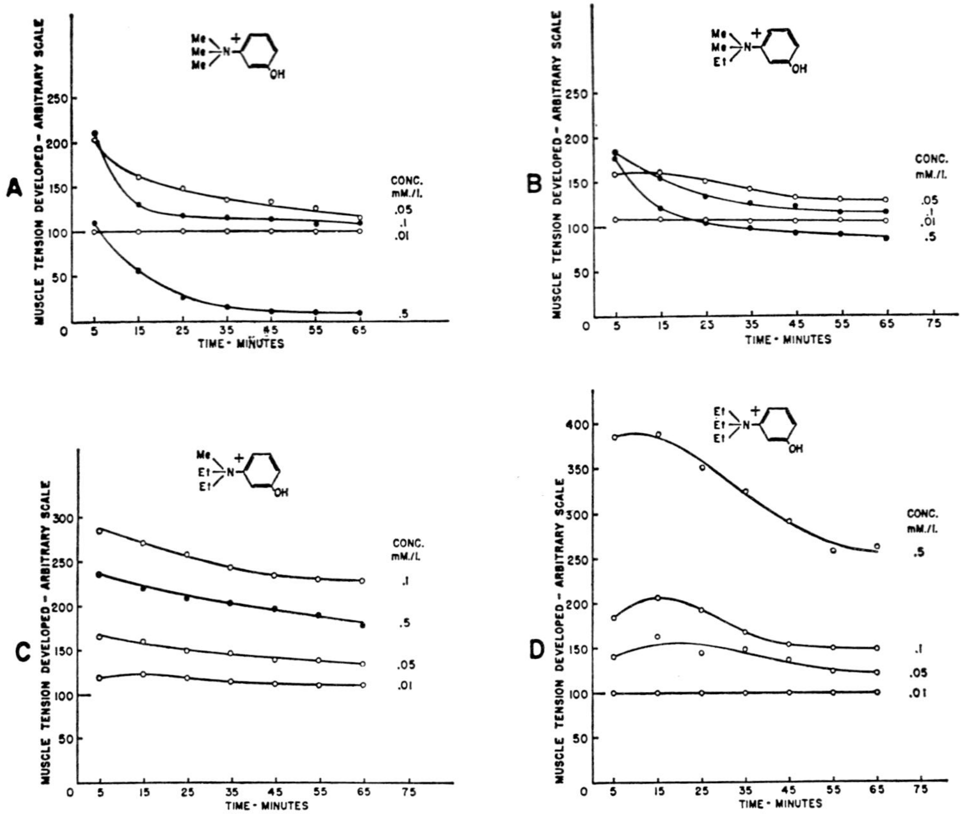

Figure 7–1 A-D. Showing maximum muscle tension output of indirectly stimulated frog sartorius muscle after application of indicated quaternary ammonium compound. For each plot, concentration of the applied drug is given at right. The nerve stimulus was a single 0.1 msec pulse. Muscle tension output during the stable control period prior to drug application equals 100 units on the arbitrary tension output scale. Open circles indicate that with progressive increase in drug concentration output tension also increases. Closed circles indicate that the output tension declined below results obtained at the next lowest concentration used. Each plot represents the average of five experiments.

logue of neostigmine was expected to have weaker direct stimulating and desensitizing action on the PJM compared with neostigmine and for this reason it might be superior to neostigmine in clinical applications.

The effect of neostigmine and diethylneostigmine (10μM) on the tension output of indirectly stimulated frog sartorious muscle *in vitro* is shown in Figure 7–2. Note that both drugs gave a well-maintained appreciable potentiation. Data from similar experiments with these drugs applied at fifty times greater concentration (500μM) show (Fig. 7–3) that the potentiation in tension output is better maintained wih diethyl-

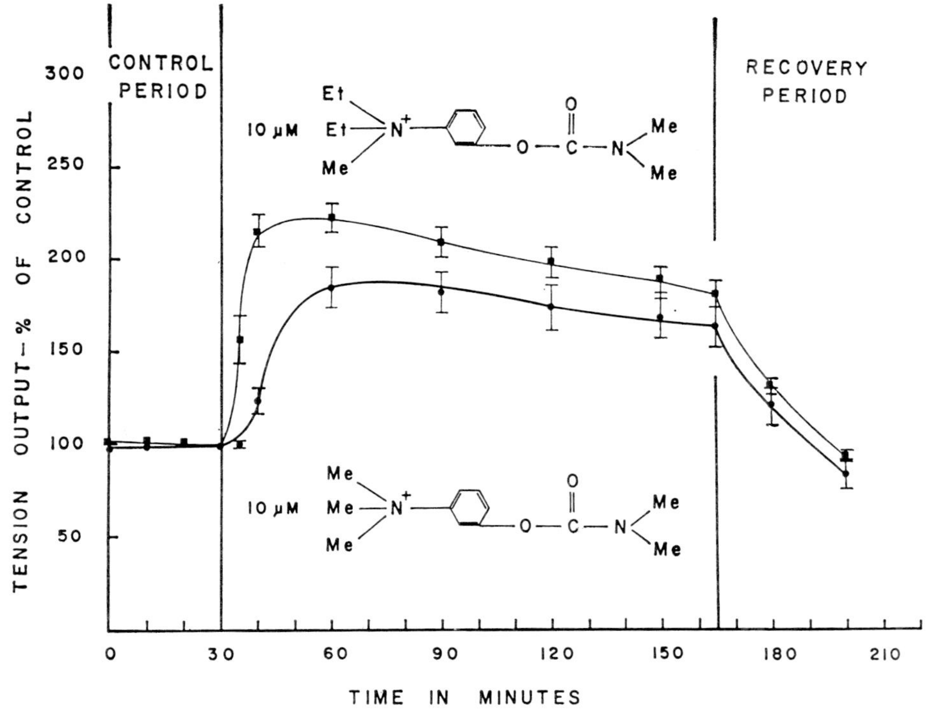

Figure 7–2. Plots showing the average tension output of indirectly stimulated frog sartorius muscles after addition of 10^{-5}M diethylneostigmine (*upper curve,* average of 14 muscles) or 10^{-5}M neostigmine (*lower curve,* average of 6 muscles). Vertical lines at each point represent the standard error of the mean.

neostigmine. At this concentration an inhibitory action of neostigmine became evident.

We next tested the capacity of these drugs to antagonize a neuromuscular block produced with d-tubocurarine. As expected, neostigmine and its diethyl analogue were both effective (Fig. 7–4) and the latter compound was slightly superior.

The next experiments were done on single fibers of the frog sartorius muscle *in vitro.* The trans-membrane potential was measured at the PJM by use of the standard intracellular recording techniques employed in our laboratory. At 5×10^{-4}M, neostigmine depolarized the PJM by 7 mV (19 cells); and at 5×10^{-3}M, the PJM depolarization was 12 mV (11 cells). At each of the above concentrations no significant PJM depolarization was produced by diethylneostigmine (19 and 28 cells). Note from Figure 7–3 that neostigmine at 5×10^{-4}M fails to maintain the potentiated tension output of indirectly stimulated muscle.

From the preceding evidence, we concluded tentatively that diethyl-

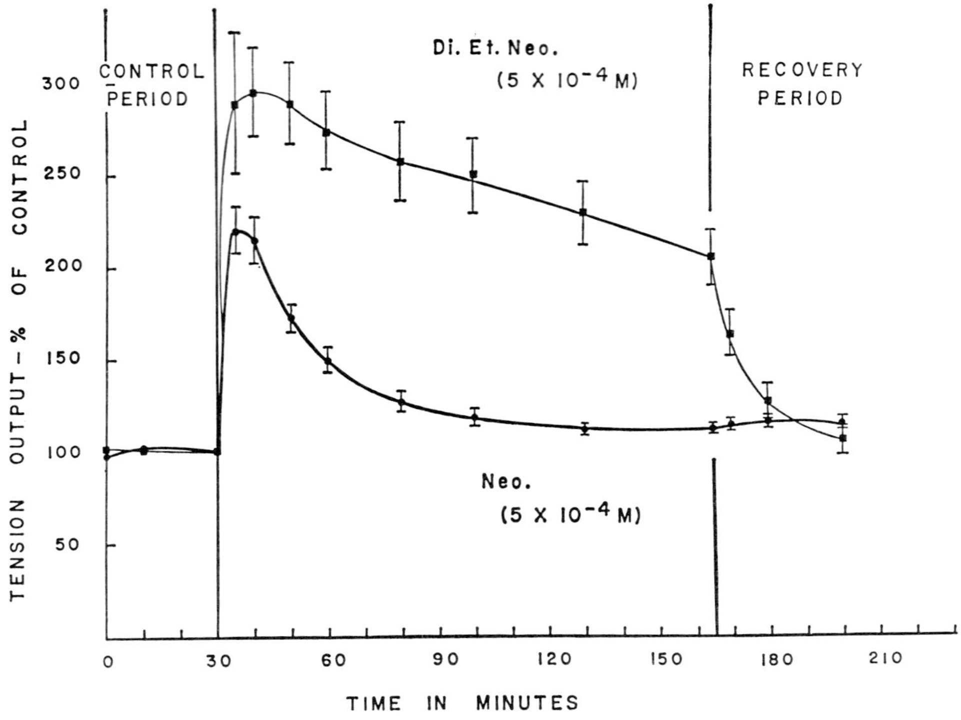

Figure 7–3. Plots showing the average tension output of indirectly stimulated frog sartorius muscles after addition of 5×10^{-4}M diethylneostigmine (*upper curve*, average of 9 muscles) or 5×10^{-4}M neostigmine (*lower curve*, average of 8 muscles). Vertical lines at each point represent the standard error of the mean.

neostigmine had relatively weak neuromuscular blocking action. However, from additional experiments, we found that the blocking capacity of this analogue was not negligible. Figure 7–5 shows resting and action potentials recorded at the PJM of indirectly stimulated muscle fibers. The record at the left was made after eighteen minutes exposure of the muscle to 5×10^{-4}M diethylneostigmine. Note the remarkably prolonged falling phase of the action potential. At the right is shown a similar record made in another fiber of the muscle but after eighty minutes exposure to diethylneostigmine. Here one can see that the endplate potential (EPP) rises relatively slowly and the falling phase of the action potential is not so greatly prolonged at low membrane potential values. These results indicate that some inhibition of neuromuscular transmission developed, although it was not sufficient to cause blockade.

Additional evidence concerning the postjunctional action of neostigmine and its diethyl analogue was obtained by study of MEPPs during

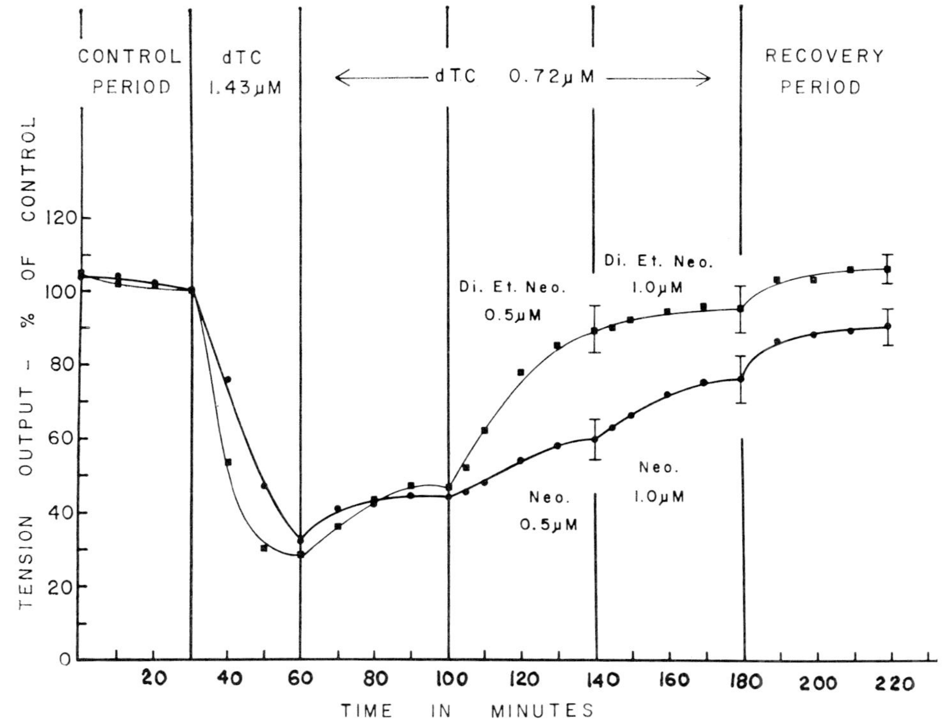

Figure 7–4. Plots showing the average tension output of indirectly stimulated frog sartorius muscles after two stage addition of d-tubocurarine. This was followed by 5×10^{-7}M diethylneostigmine (*light line,* average of 6 muscles) or 5×10^{-7}M neostigmine (*heavy line,* average of 5 muscles). Thereafter the drug concentration was increased to 1×10^{-6}M. Vertical bars represent the standard error of the mean.

perfusion of the neuromuscular junction with Ringer's solution containing neostigmine 1×10^{-4}M. During such perfusion the amplitude of the MEPPs at first increased and then decreased. If the concentration of applied neostigmine was increased to 2×10^{-3}M, the MEPPs simply decreased in amplitude and often became too small to be detected. Similar inhibition of MEPPs was obtained by perfusing junctions with diethylneostigmine at ten times lower concentrations.

The potentiation in tension output of the indirectly stimulated muscle exposed to neostigmine and its diethyl analogue results from the fact that under such treatment a single nerve volley initiates not one, but a train of action potentials in each muscle fiber. Thus a single neuronal action potential (which would normally evoke a simple twitch in a control preparation) now produces a brief tetanus. An example of such repetitive firing produced by a single neuronal stimulus is shown in Figure 7–6. The preparation was pretreated with 1×10^{-5}M neostigmine.

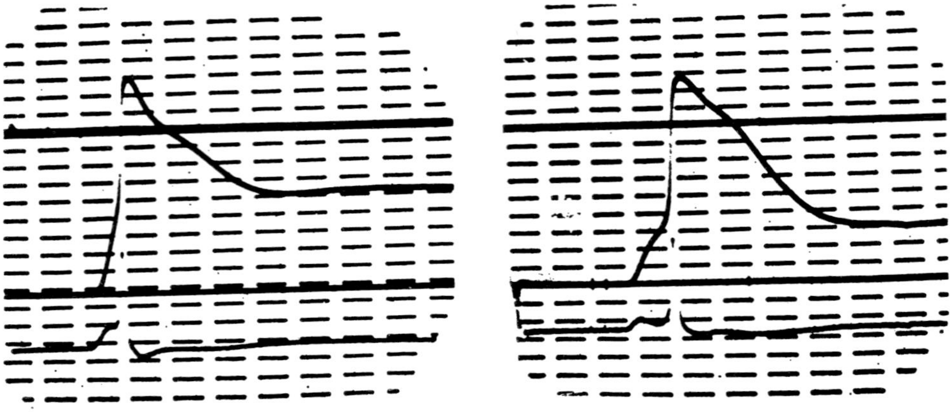

Figure 7–5. Action potential (*left*) recorded at the junctional region of a muscle fiber exposed for eighteen minutes to 5×10^{-4}M diethylneostigmine. At the right is a similar recording obtained after eighty minutes' exposure to this drug. Calibration grid: Y axis, 10 mV steps, X axis, 1 msec intervals. Lower trace is the time derivative of the action potential for which each Y axis calibration step represents 100 V/sec.

The action of neostigmine (5×10^{-5}M) in increasing the amplitude of endplate potentials (EPP) of muscle fibers in which neuromuscular transmission was blocked by application of d-tubocurarine (7×10^{-6}M) is shown in Figure 7–7. The transmembrane potential was recorded in the usual manner, and the motor nerve was stimulated at five-second intervals to evoke EPPs. On the left are records obtained with a slow

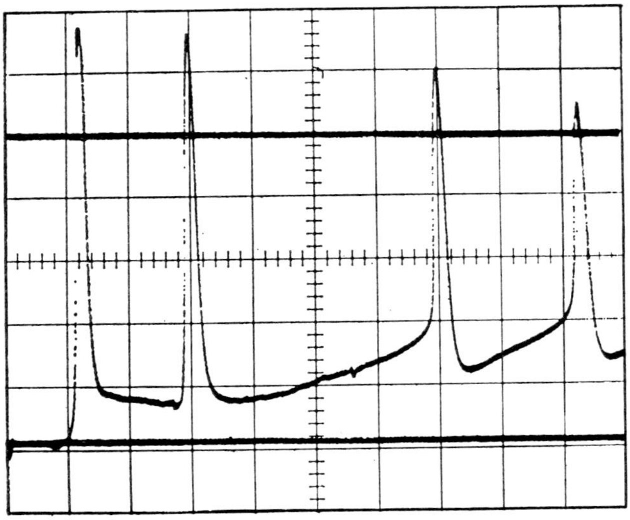

Figure 7–6. Multiple action potentials recorded from a single muscle fiber at a non-junctional region. The muscle was equilibrated in neostigmine 1×10^{-5}M and a single supramaximal 0.1 msec stimulus was applied to the motor nerve. Y axis calibration: 20 mV/block; X axis calibration: 5 msec/block.

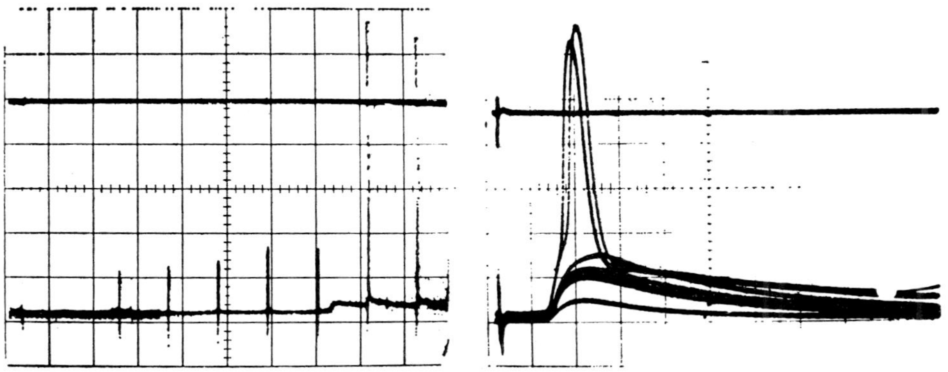

Figure 7–7. Showing (*left*) the increase in amplitude of endplate potentials recorded from a single curarized fiber during perfusion of the junctional region with neostigmine (5×10^{-5}M). The nerve was stimulated at 5 sec intervals with 0.1 msec stimuli. The endplate potentials evoked by the last two stimuli were sufficient in amplitude to initiate action potentials. Y axis = 20 mV/block; X axis = 5 sec/block. *Right*, Same recordings but taken with overlapping sweeps. Y axis = 20 mV/block; X axis = 2 msec/block.

sweep showing that application of neostigmine rapidly increased the EPP amplitude and finally led to initiation of an action potential. A composite oscillogram of these records taken at faster sweep speed is shown at the right.

We assume that the facilitatory action of neostigmine and its diethyl analogue on neuromuscular transmission depends principally on the powerful anticholinesterase activity which they exert at postjunctional regions. The basic mechanism underlying such facilitation has been well described in the literature (e.g. reference 8). Our present investigations were designed to explore adjunct properties of these drugs which do not depend on acetylcholinesterase inhibition, i.e. their direct stimulating or blocking action on the PJM. To these we added a brief exploration of the possible action of neostigmine on motor nerve terminals, using the experimental arrangement shown in Figure 7–8 A. In these experiments the nerve-muscle preparation was first equilibrated in Ringer's solution containing d-tubocurarine 3μM to 7μM. This was done to block neuromuscular transmission in order to eliminate recording artifacts produced by muscle contraction. A recording electrode (1μm tip dia.) filled with Ringer's solution was placed extracellularly on one of the neuronal filaments of the motor terminal arbor for recording the nerve action potential during its conduction into these terminal elements. Finally, a pipette (50μm tip dia.) filled with Ringer's solution containing d-tubocurarine (3μM–7μM) and neostigmine 1.3×10^{-5}M was placed near the neuromuscular junction in order to perfuse this region. Some typ-

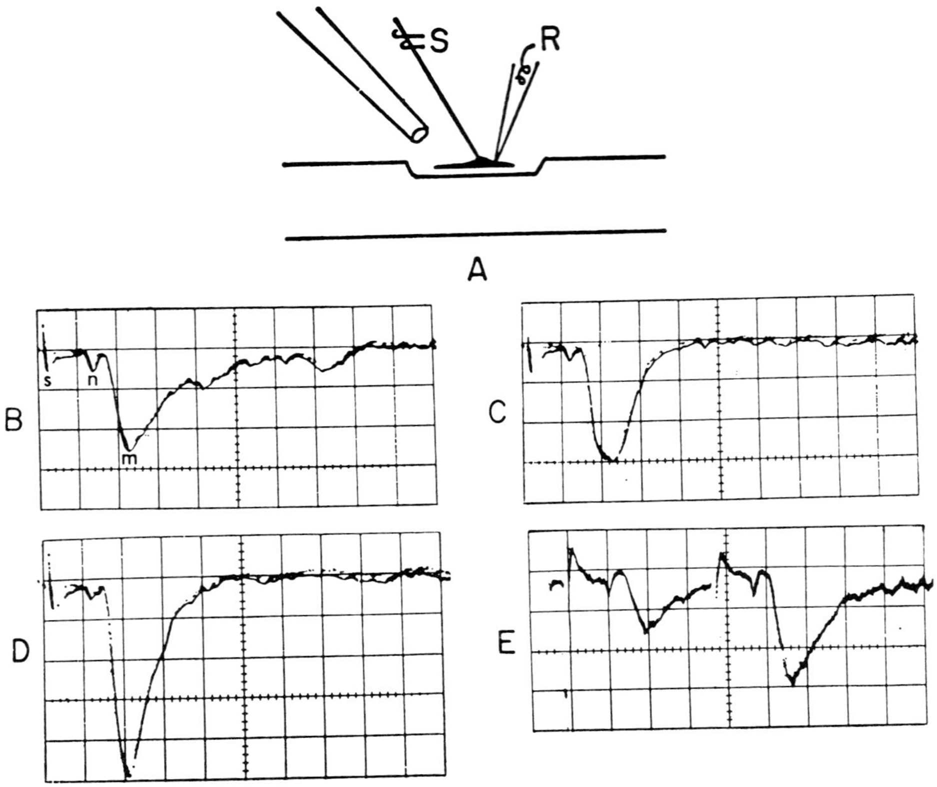

Figure 7–8 A-E. *A*, Model of the experiment in which extracellular recordings (R) are made at a terminal filament of the motor neuron. S = nerve-stimulating electrodes, 0.1 msec pulses. The muscle was bathed in Ringer's solution containing d-tubocurarine (3 to 7 × 10^{-6}M). *B*, Control record in which the stimulus artifact (s) is followed by the neuronal action potential (n) which in turn is immediately followed by an endplate potential (m). *C* and *D*, Similar records obtained after microperfusion of the junction with Ringer's solution containing d-tubocurarine (3 to 7 × 10^{-6}M) and neostigmine 1.3 × 10^{-5}M. *E*, Record made as in *B* but with two nerve stimuli 8 msec apart. X axis = 2 msec/block.

ical oscillograms are shown in Figure 7–8 B-E. The control record (Fig. 7–8 B) shows the stimulus artifact followed by the neuronal action potential, which in turn is immediately followed by the EPP. Records shown at Figure 7–8 C and D were obtained during perfusion of the junctional area with neostigmine. Note the increased EPP amplitude and the unchanged neuronal action potential. No evidence of repetitive neuronal firing was obtained. The record at Figure 7–8 E shows the response to two nerve stimuli given in rapid succession to indicate that our recording technique is capable of detecting two neuronal impulses originating with small time delay.

From the above experiments we concluded that drugs such as neostigmine have some capacity to depolarize and desensitize the PJM and by such means might produce some of the disturbances seen in "cholinergic crisis." Diethylneostigmine also desensitizes the PJM but without depolarizing it. These direct actions are produced when relatively high concentrations of neostigmine and diethylneostigmine are applied. It is more difficult to assess their importance under *in vivo* conditions where one usually encounters repetitive activity, not single isolated discharges at the neuromuscular junction. We judge, nonetheless, that our data provide some support for suggesting that the diethyl analogue of neostigmine might be a better drug than neostigmine in aiding patients with myasthenia gravis.

We now turn to another group of pharmacological agents which have been used in the treatment of myasthenia gravis: the veratrum alkaloids. When compounds of this type are applied to nerve or muscle fibers they cause singly stimulated cells to fire repetitively. Thus, for example, application of a short enduring electrical stimulus evokes not one action potential but a long train of action potentials. The possible use of the veratrum alkaloids in myasthenia gravis has the following rationale. In this disease the patient has difficulty maintaining muscle tension in a tetanus. Although the muscle tension can reach large values at the start of a volitional act, the tension output falls rapidly if the patient attempts to sustain muscle contraction. Suppose that after treatment with a veratrum alkaloid each individual nerve impulse initiated not the usual single muscle impulse but a train of muscle impulses which generate a muscle tetanus. Under such conditions each muscle tetanus could be evoked at low physiological cost for the presynaptic elements and other processes involved in neuromuscular transmission and these elements could be spared from fatigue. In order to achieve this theoretically attractive relationship between presynaptic and postsynaptic elements, the veratrum alkaloid would have to be applied in a critical concentration in order to produce repetitive firing in muscle without causing repetitive firing in nerve. One of our goals in this investigation was to determine whether one can achieve such relative selective action.

Among the veratrum alkaloids, germine monoacetate (the active component of germine diacetate) has been considered to be potentially useful for clinical application because it is relatively free of undesirable side effects. Evidence for this statement as well as other important information concerning germine diacetate and germine monoacetate can be found in several papers.[9-14]

Our own presently reported experiments were conducted with germine diacetate (GDA) which was kindly supplied to us by Dr. Francis

F. Foldes. At the time we began our work, we (and others) were not aware of the fact that the germine diacetate used also contained germine and germine monoacetate. Apparently germine monoacetate is the most active of the three compounds, but all appear to have similar pharmacological properties. We therefore believe it worthwhile to report the results of our earlier experiments which were done on the frog sciatic nerve-sartorius muscle preparation *in vitro*. Following Flacke's method, GDA 1 gm/liter was dissolved in distilled water which contained HCl in an equimolar amount. This stock solution was stored at 4°C. All Ringer's solutions containing GDA were freshly prepared from this stock solution and were used in six hours or less. Bath temperatures ranged from 18°–20°C.

In the first series of experiments the nerve-muscle preparation was bathed in Ringer's solution containing GDA (5 mg/liter) and d-tubocurarine (25 mg/liter). After thirty minutes exposure to this solution a single nerve stimulus produced only a single EPP without any signs of repetitive activity and the preparation did not twitch. A diagram of the experimental arrangement is shown in Figure 7–9 A. We removed d-tubocurarine(dTC) by perfusing the neuromuscular junction with Ringer's solution containing only GDA 5 mg/liter as shown. Double 0.1 msec stimuli about 10 msec apart were applied to the nerve to promote repetitive firing. Figure 7–9 B shows the results obtained after three minutes of perfusion with dTC free solution. In this record the middle trace, which represents the muscle transmembrane potential record, shows in succession the first stimulus artifact, two succeeding EPPs, the second stimulus artifact, and a following train of EPPs appearing at approximately 5 msec intervals. Figure 7–9 C shows the same junction after five minutes of perfusion. The EPPs have increased in amplitude; and after the second nerve stimulus was applied, a train of EPPs was initiated. Alternate EPPs of this train initiated muscle action potentials. In a group of such experiments we always obtained repetitive neuronal firing (as judged by trains of EPPs produced). This result is taken to mean that GDA effects repetitive neuronal firing more readily than repetitive firing in the muscle fiber. In connection with the interpretation of these experiments it may be noted that dTC even in high concentrations does not prevent applied veratrum alkaloids from causing repetitive firing in muscle fibers.[15]

The next experiments were carried out to determine the locus of the repetitive firing produced by GDA. Figure 7–10 A shows the experimental arrangement. Electrodes (S-N) placed on the nerve supplied neuronal stimuli. One intracellular electrode (R-M) located approximately 2 mm from the neuromuscular junction was used to record muscle action potentials. Via a second intracellular electrode (S-M)

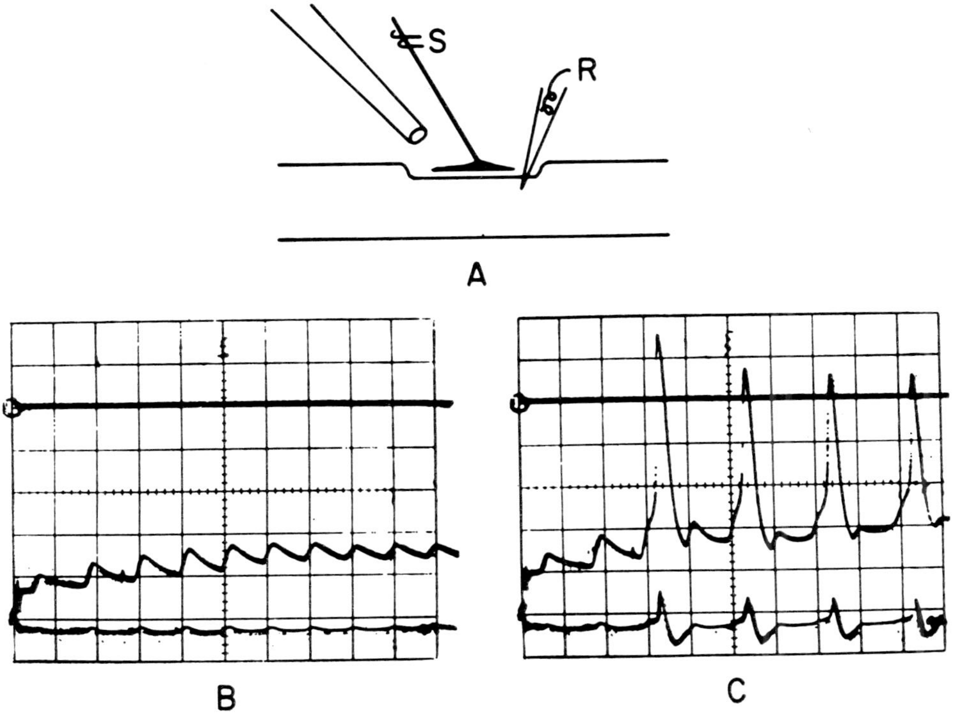

Figure 7–9 A-C. *A,* Model of the experiment. R = membrane potential recording electrode. S = stimulating electrodes. Bath contains Ringer's solution plus GDA 5 mg/liter plus d-tubocurarine 25 mg/liter. The perfusion pipette contains Ringer's solution plus GDA (5 mg/liter). *B,* Middle trace, repetitive endplate potentials produced by dual nerve stimulation. *C,* Middle trace, repetitive endplate potentials and initiated action potentials produced by dual nerve stimulation. Y axis = mV/block; X axis = 5 msec/block. Lower traces in *B* and *C* represent rate of change of membrane potential with time.

placed nearby, brief (0.1 – 0.2 msec) electrical stimuli were applied directly to the muscle fiber. The preparation was bathed in Ringer's solution containing GDA 10 mg/liter. Figure 7–10 B, which is a record made with a fast sweep, shows the response of the muscle fiber to a single direct stimulus. One can see at the left the large stimulus artifact followed immediately by a muscle action potential which was initiated and recorded at the site of stimulation. Note the prominent negative after-potential. Figure 7–10 C shows the same record as in Figure 7–10 B but taken at a ten times slower sweep speed. At the left hand edge of this record one can see only the last portion of the falling phase of the first muscle action potential, but after a latency of about 25 msec a train of muscle action potentials appears. We interpret these records to mean that during direct stimulation of the muscle fiber, an action potential is initiated at the site of stimulation. Thereafter this action potential

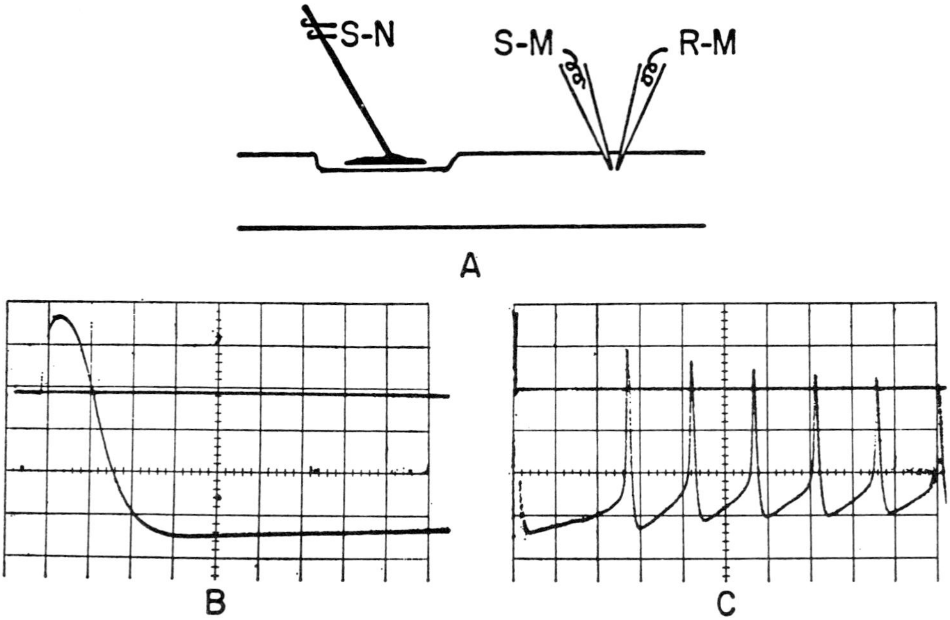

Figure 7–10 A-C. *A*, Model of experiment in which S-N = nerve-stimulating electrode. S-M = muscle fiber intracellular stimulating electrode. R-M = muscle fiber intracellular recording electrode. Bath contains Ringer's solution plus GDA 10 mg/liter. *B*, Muscle action potential initiated by stimulating the muscle fiber directly. Y axis = 20 mV/block; X axis = 1 msec/block. *C*, Same record as *B* but made at slower sweep speed. Y axis = 20 mV/block; X axis = 10 msec/block.

travels along the fiber until it reaches some region distant from the site of stimulation, whereupon a focus of repetitive activity is set up and successive muscle action potentials are initiated which are propagated back to the recording point. We wondered whether such foci of repetitive firing occur at random along the length of the muscle fiber or whether they might have some preferential location, say at the neuromuscular junction. Our next experiments were designed to answer this question. Figure 7–11 A shows the experimental arrangement. Intracellular records were made at points equidistant (0.3 mm) from the neuromuscular junction (R-M1 and R-M2), and the muscle fiber was directly stimulated very close to the recording electrode R-M1. When a *single* stimulus was applied, the first action potential was recorded at R-M1 with practically no latency and it was recorded at R-M2 0.3 msec later (Fig. 7–11 B). This is expected because of the additional time consumed in conduction of the action potential from right to left along the muscle fiber. Thereafter, however, the next action potentials which arrived at the R-M1 and R-M2 recording electrodes appeared in almost perfect synchrony (Fig. 7–11 C). We inferred from these facts that in its wake

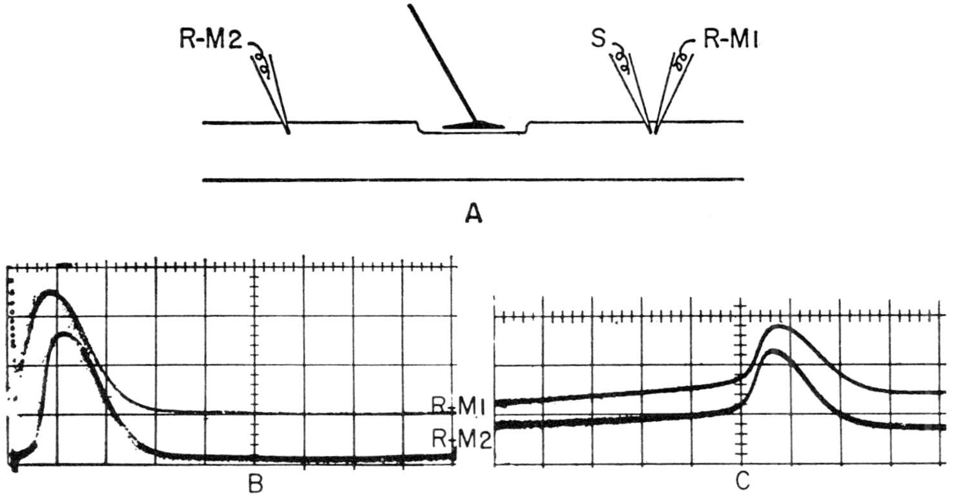

Figure 7–11 A-C. *A*, Model of experiment in which S = intracellular muscle stimulating electrode, and R-M1 and R-M2 = intracellular muscle fiber recording electrodes placed equidistant from the neuromuscular junction. Bath contains Ringer's solution plus GDA 10 mg/liter. *B, Upper trace,* Muscle action potential recorded at R-M1 in response to single stimulation at S. *Lower trace,* Muscle action potential recorded at R-M2 approximately 0.3 msec later. Y axis = 25 mV/block; X axis = 1 msec/block. *C, Upper trace,* Second muscle action potential recorded at R-M1. *Lower trace,* Second muscle action potential recorded at R-M2. Y axis = 25 mV/block; X axis = 1 msec/block.

the first initiated muscle action potential set up a site of repetitive firing and that this site was equidistant from the recording electrodes, i.e. it was located at the neuromuscular junction. Evidently muscle action potentials making a transit through the junctional region can cause repetitive firing of some structure located therein, but one does not know whether this element is the neuronal terminals overlying the PJM, the PJM itself, or perhaps the muscle fiber membrane contiguous with the PJM.

Further evidence bearing on this matter was supplied by the next experiments. The experimental arrangement used was the same as that shown in Figure 7–10 A. The transmembrane potential record was taken some distance from the neuromuscular junction but in this case a single stimulus was applied to the *nerve.* Note in Figure 7–12 (*left*) taken with a fast sweep, that the first muscle action potential appears at the recording point with measurable latency. This latent period represents neuronal conduction time, neuromuscular transmission delay, and the conduction time for the muscle action potential to travel from the neuromuscular junction to the recording point. Note in Figure 7–12

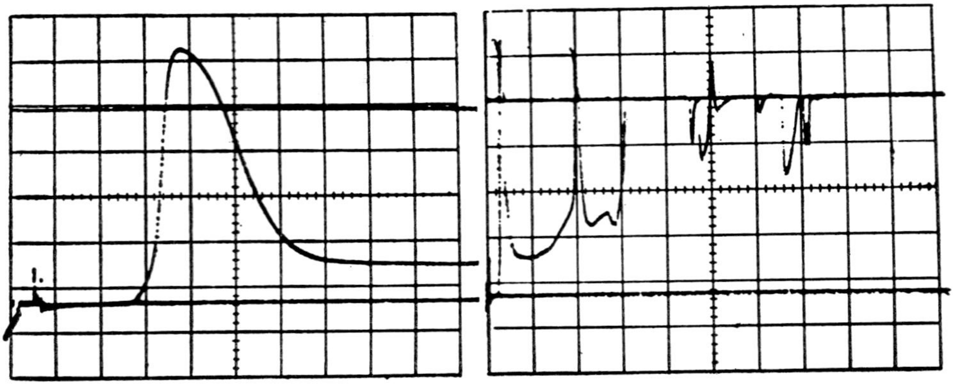

Figure 7–12. Same experimental arrangement as in Figure 7–10A. Bath contains Ringer's solution plus GDA 10 mg/liter. *Left record,* Neurally evoked muscle action potential recorded some distance from the neuromuscular junction. Y axis = 20 mV/block; X axis = 1 msec/block. *Right record,* Same record as above but taken at slower sweep speed. Y axis = 20 mV/block; X axis = 10 msec/block.

(*right*), which is a record taken at a ten times slower sweep speed, that there is substantial latency between the first and second muscle action potentials recorded. Unfortunately the strong contraction of the entire muscle produced by motor nerve stimulation prevented us from recording succeeding muscle action potentials in the train. We infer from the results of this experiment and those preceding that repetitive firing is initiated at the neuromuscular junction with one of the junctional elements acting as the initiator of successive muscle action potentials.

In the final set of experiments, we further attempted to localize the locus of repetitive firing of neurally stimulated muscle fibers. The experimental arrangement is shown in Figure 7–13 A. The muscle was bathed in Ringer's solution containing dTC 1.5 mg/liter. One recording electrode (R-N) was placed extracellularly on a motor neuron terminal; the second electrode (R-M) was placed intracellularly for recording PJM potentials. A perfusion pipette loaded with Ringer's solution containing dTC 1.5 mg/liter and GDA 10 mg/liter was positioned at the junctional region at an appropriate time in order to apply GDA. Figure 7–13 B-C shows control records taken in the absence of GDA. The upper traces (R-N) are extracellular records obtained from electrode R-N located on the neuron terminal. In Figure 7–13 B, taken at slow sweep speeds, or in more detail in Figure 7–13 C, which represents the same records taken at five times greater sweep speed, one can see the nerve stimulus artifact (s) followed by the neuronal action potential (n) and the EPP (m). The lower traces of Figure 7–13 B-C show the EPP at slow and fast sweep speeds as recorded from electrode R-M in the

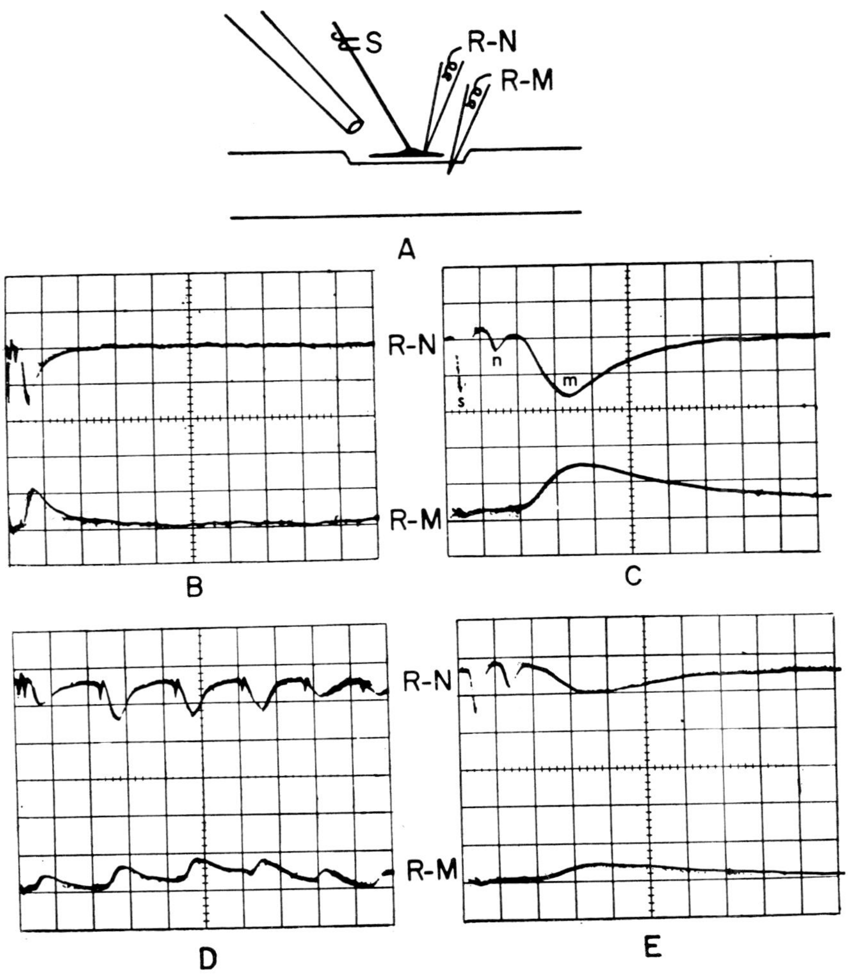

Figure 7–13 A-E. *A,* Model of the experiment. S = nerve-stimulating electrodes. R-N = extracellular recording electrode on motor neuron terminal. R-M = muscle fiber intracellular recording electrode. Bath contains Ringer's solution plus d-tubocurarine 1.5 mg/liter. Perfusion micropipette contains Ringer's solution plus d-tubocurarine 1.5 mg/liter plus GDA 10 mg/liter. *B,* Control experiment (no GDA applied). *Upper trace,* Neuronal action potential recorded at R-N. Y axis = 0.1 mV/block. *Lower trace,* Endplate potential recorded at R-M. Y axis = 0.5 mV/block; X axis = 5 msec/block for both traces. *C,* Same as *B* but with sweep speed increased to 1 msec/block. Recording made at R-N shows s = stimulus artifact, n = neuron action potential, m = endplate potential. *D,* Recording made as in *B* but after ten minutes' perfusion of the junction with solution containing GDA. Note repetitive firing produced by single nerve stimulus. E, Same as *D* but with sweep speed increased to 1 msec/block.

Chapter 8

The Role of Enzymes in Carbohydrate Metabolism and Hemorrhagic Shock

OLGA SCHWEIZER

Intermediary carbohydrate metabolism is intimately related to the biochemical changes associated with hemorrhagic shock. In both conditions enzyme and coenzyme systems play a fundamental role.

The end result of all oxidative processes is the union of hydrogen or electrons with oxygen to form water. In carbohydrate metabolism the various steps preceding the ultimate formation of water are divided into two main phases. The first, or anaerobic, phase is known as glycolysis and can proceed in the absence of oxygen. The second phase, which requires oxygen, is designated by several names: Krebs cycle, citric acid cycle, or tricarboxylic acid cycle. The enzymes and coenzymes involved in the aerobic Krebs cycle are located within the mitochondria, whereas those concerned wtih the glycolytic process are restricted to the extramitochondrial portions of the cell (the cell sap).

Figure 8–1 shows the major steps in glycolysis with their respective catalytic enzymes. Each mole of glucose yields 2 moles of pyruvate. The reaction catalyzed by glyceraldehyde 3-P dehydrogenase is of particular interest, since it involves the reduction of oxidized nicotinamide adenine dinucleotide (NAD^+) to NADH. Under normal aerobic conditions, the reoxidation of this extramitochondrial NADH takes place within the mitochondria via the electron transport chain. Since the mitochondrial membrane is impermeable to the NADH of the cell sap, indirect methods or shuttle systems must be employed to transfer reducing equivalents across the membrane. Two of these systems, the glycerophosphate shuttle and the acetoacetate B-hydroxybutyrate shuttle are illustrated in Figure 8–2. In the glycerophosphate shuttle, dihydroxyacetone phosphate is converted to α-glycerophosphate with the concomitant reoxidation of NADH, according to the following reactions:

$$\text{glyceraldehyde 3-phosphate} + P_i + NAD^+ \xrightleftharpoons[]{\left[\begin{array}{c}\text{glyceraldehyde 3-P}\\ \text{dehydrogenase}\end{array}\right]} \text{1, 3-diphosphoglycerate} + NADH + H^+$$

$$NADH + H^+ + \text{dihydroxyacetone phosphate} \leftrightarrows NAD^+ + \alpha\text{-glycerophosphate}$$

From the Department of Anesthesiology, Memorial Sloan-Kettering Cancer Center, New York, N.Y. 10021.

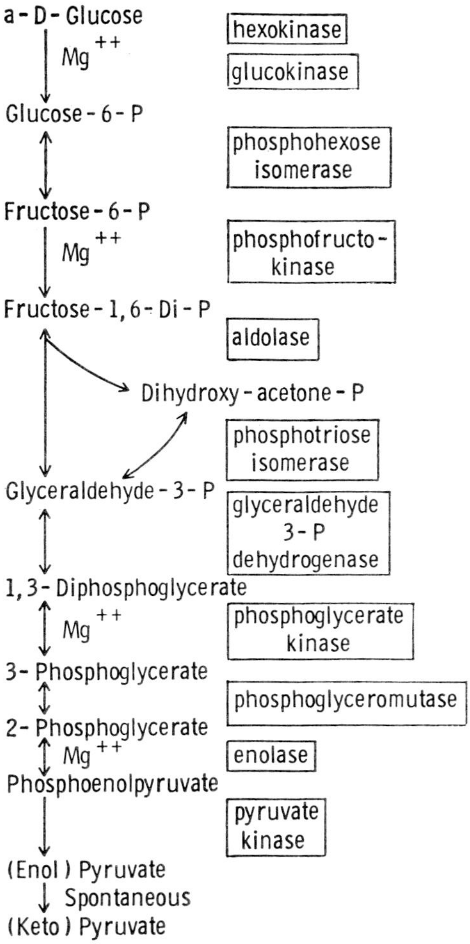

Figure 8–1. Glycolysis.

In contrast to NADH the α-glycerophosphate readily penetrates the mitochondrial membrane, where it is reoxidized via the intramitochondrial electron transport chain. A similar reaction occurs in the B-hydroxybutyrate shuttle, in which extramitochondrial acetoacetate accepts electrons from NADH to form B-hydroxybutyrate. Within the mitochondria, the B-hydroxybutyrate is reoxidized to acetoacetate.

The oxidation of pyruvate (the end product of glycolysis) to carbon dioxide and water is completed in the Krebs cycle (Fig. 8–3). Pyruvate is converted to acetyl-Co A which combines with oxaloacetate to form citrate. The cycle then continues through cis-aconitate, isocitrate, oxalosuccinate, α-ketoglutarate, succinyl-Co A, succinate, fumarate, malate, and back to oxaloacetate. At five sites in the cycle (reactions catalyzed

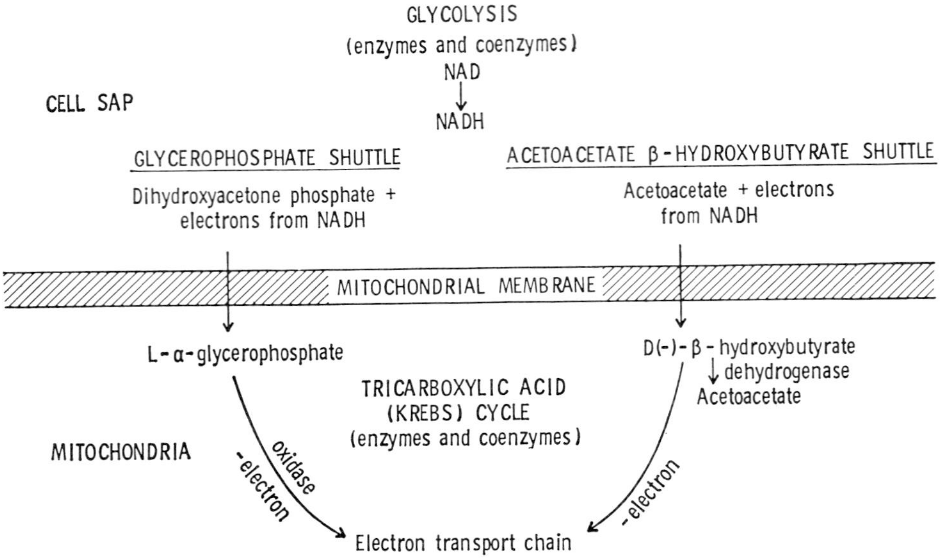

Figure 8–2. Shuttle systems.

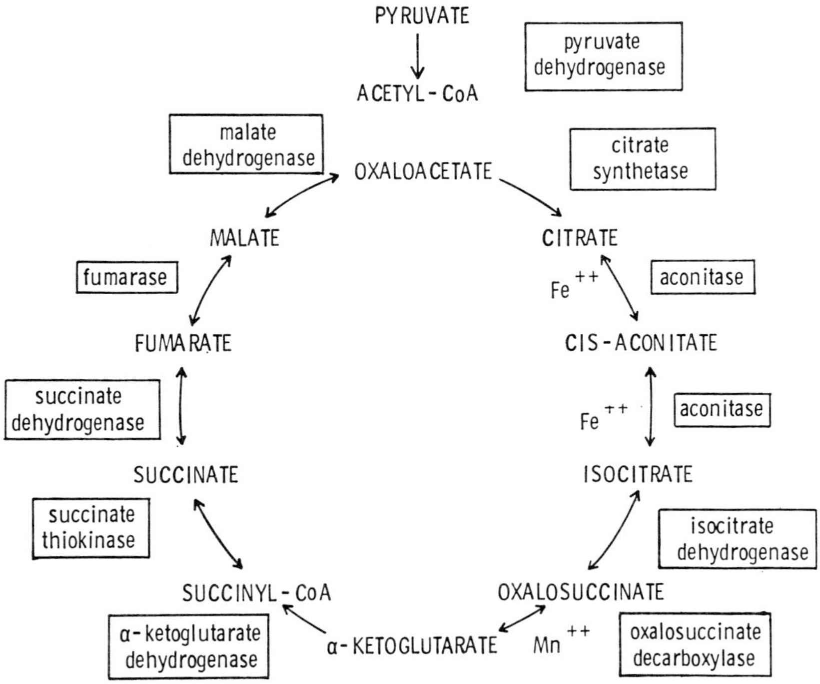

Figure 8–3. Citric acid (Krebs) cycle.

by pyruvate dehydrogenase, isocitrate dehydrogenase, α-ketoglutarate dehydrogenase, succinate dehydrogenase and malate dehydrogenase) hydrogen or electrons are transferred from the substrate acids to the intramitochondrial electron transport chain for final union with oxygen to form water.

The intramitochondrial electron transport chain consists of several compounds with one property in common: the capability of alternate oxidation and reduction. By this means hydrogen ions and electrons are transferred along the chain to the terminal oxygen. The components of the chain (Fig. 8–4), in order, include nicotinamide adenine dinucleo-

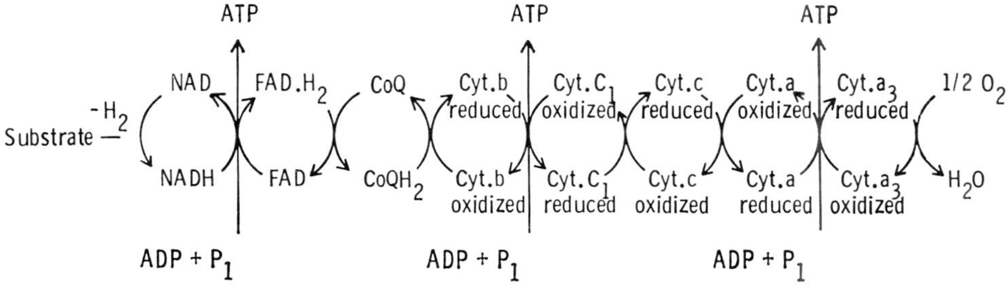

Figure 8–4. Intramitochondrial electron transport chain.

tide (NAD), flavoproteins (FAD), coenzyme Q (CoQ), and cytochromes b, c_1, c, a and a_3. In four of the five Krebs cycle reactions catalyzed by dehydrogenases, electrons are transferred from the substrate to the first component of the chain, the coenzyme NAD. The reaction catalyzed by succinate dehydrogenase differs from the others in that the electrons are transferred directly to the flavoproteins, bypassing NAD.

The oxidative processes of the Krebs cycle result in the release of energy. According to Schatz,[1] this energy is first converted into a high-energy intermediate compound, the exact nature of which has not been determined at the present time. The energy of the intermediate compound is then diverted into three pathways: (1) the formation of adenosine triphosphate (ATP), (2) a source of energy other than ATP for ion-transport, biosyntheses and similar processes, and (3) dissipation as heat. Lehninger[2] states that approximately 50 per cent is preserved for future use as the high-energy compound, ATP.

The formation of ATP by phosphorylation of ADP is closely coupled with the intramitochondrial electron transport chain at three sites (Fig. 8–4). This means that six high-energy phosphate bonds are formed every time two electrons traverse the entire chain from NAD to oxygen. In the case of succinate, the formation of high-energy bonds is limited to four, since the reaction bypasses the first phosphorylation site between

NAD and the flavoproteins. Substrate oxidation accounts for two additional ATP, making a total of 30 moles of ATP produced during the aerobic Krebs cycle.

Although the major part of ATP formation occurs in association with the intramitochondrial electron transport chain, Table 8-I shows that high-energy bond production is not limited to the aerobic phase of carbohydrate metabolism. Four bonds are formed during glycolysis as a result of substrate oxidation and six in the reaction catalyzed by glyceraldehyde-3-phosphate dehydrogenase. Since glycolysis utilizes two

TABLE 8–I
FORMATION OF HIGH ENERGY BONDS

	Enzyme Catalyst	*Method of Bond Production*		*No. of Bonds per Mol Glucose*
GLYCOLYSIS	Glyceraldehyde-3-phosphate dehydrogenase	Respiratory chain-2NADH		6
	Phosphoglycerate kinase	Substrate oxidation		2
	Pyruvate kinase	Substrate oxidation		2
				10
	Utilized during glycolysis			−2
			Net	8
CITRIC ACID CYCLE	Pyruvate dehydrogenase	Respiratory chain-2NADH		6
	Isocitrate dehydrogenase	Respiratory chain-2NADH		6
	α-Ketoglutarate dehydrogenase	Respiratory chain-2NADH		6
	Succinate thiokinase	Substrate oxidation		2
	Succinate dehydrogenase	Respiratory chain-2$FADH_2$		4
	Malate dehydrogenase	Respiratory chain-2NADH		6
			Net	30

ATP, the end result is a net production of eight high-energy bonds. The total high-energy yield for both phases of carbohydrate metabolism is 38 moles of ATP for each mole of glucose.

A variety of conditions can produce marked alterations in the normal aerobic metabolic process. These include inhibition of any of the components of the intramitochondrial electron transport chain, blockage of the shuttle systems for transfer of reducing equivalents across the mitochondrial membrane (certain types of cancer cells), lack of oxygen in the inspired air, and decreased tissue perfusion. In any of these situations the usual intramitochondrial pathway for reoxidation of extramitochondrial NADH is no longer available. Since oxidized NAD^+ is essential for the continuation of the glycolytic process, an alternative method confined to the extramitochondrial portion of the cell must be provided. This method involves the reduction of pyruvate to lactate with the concomitant reoxidation of NADH to NAD^+ (Fig. 8–5). The result is a marked increase in total and excess lactate production.

Restriction of carbohydrate metabolism to the anaerobic phase also involves a considerable reduction in the formation of high energy phosphate compounds. Since the 30 moles of ATP of the Krebs cycle are formed in association with the intramitochondrial electron transport chain, the conversion to anaerobic respiration eliminates thirty of the available thirty-eight high-energy bonds produced during the aerobic metabolism of a mole of glucose. The dependence of anaerobic ATP production on the reduction of pyruvate to lactate necessitates the metabolism of greatly increased amounts of glucose to provide the energy necessary for cellular functions. In addition to the loss of high-energy phosphate associated with carbohydrate metabolism, the shift from aerobic to anaerobic respiration also eliminates ATP formed during the final Krebs cycle metabolism of fats and proteins.

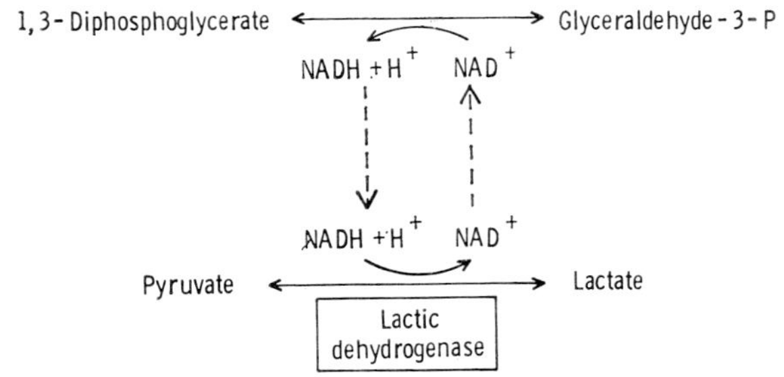

Figure 8–5. Anaerobic oxidation.

Since hemorrhagic shock is associated with cellular hypoxia secondary to decreased tissue perfusion, a shift from aerobic to anaerobic carbohydrate metabolism could be anticipated. Howland and associates[3] conducted a series of animal investigations to determine the extent of this shift. In these studies, one hundred and fifty dogs of both sexes with an average weight of 18.2 kg were splenectomized seven to fourteen days prior to the induction of hemorrhagic shock. One to two weeks after splenectomy the dogs were anesthetized with intravenous pentobarbital (30 mg/kg) and intubated. The inspired medium was room air, and spontaneous respirations were not assisted at any time. Bardic® extra large catheters were inserted into the exposed right femoral artery and vein. After control parameters and arterial blood samples had been obtained, the dogs were phlebotomized via the femoral artery to an end point of 50 mm Hg blood pressure. This required an average of 2.5 minutes. The blood was collected in an Abbott blood collection bottle containing 5 mg of heparin. Repeat samples and parameters were determined at the end of a one-hour period of stable 50 mm Hg pressure. Additional blood was then removed to an end point of 30 mm Hg blood pressure, which was maintained for forty-five minutes. Final samples and parameters were obtained at the end of this second shock period.

The dogs were monitored for blood pressure, pulse rate, cardiac output, electrocardiographic tracing, respiratory rate, and minute volume respiration. The blood pressure was recorded by means of an intra-arterial catheter attached to a Honeywell multitrace oscilloscope with a statham strain gauge + P23AA. Cardiac output measurements were determined by a Lexington cardiac output computer with a Gilson densitometer, a GME dye tracer and direct reading meter with a Harvard constant withdrawal pump. A Corbin Farnsworth scopette recorded the electrocardiographic tracings. Pulmonary ventilation was measured by the Wright Respirometer (British Optical).

In addition to the above parameters, arterial blood samples were analyzed for hematocrit, oxygen tension, pH, carbon dioxide tension, total carbon dioxide, glucose, pyruvate, lactate, excess lactate, citrate, α-ketoglutarate, malate, inorganic phosphate, adenosine monophosphate (AMP), adenosine diphosphate (ADP), and adenosine triphosphate (ATP). Values for oxidized (NAD^+) and reduced (NADH) nicotinamide adenine dinucleotide were obtanied in a second group of six dogs subjected to the same experimental conditions described for the main series. In this group samples were taken in the control period and at the end of the second shock period. A Clay Adams microcentrifuge was employed for hematocrit readings utilizing a microcapillary method. Acid-base parameters and PaO_2 were measured by an I.L. pH gas analyzer, Model 113-S_2, with pH electrode assembly #11070, a pCO_2 combination pH reference electrode with Lucite® jacket and a PO_2 electrode. The method described by Pincus, modified by Elliott,[4] was employed for citrate values. Pyruvate, lactate, α-ketoglutarate, malate, AMP, ADP, ATP, NAD^+, and NADH values were determined by enzymatic methods.[5] Excess lactate was calculated from Huckabee's formula.[6]

The values of the various parameters measured are shown in Tables 8-II, 8-III, 8-IV, 8-V, and 8-VI. The tables record only the mean ± standard deviation of the control levels and the values at the end of the second shock period. The levels at the termination of the first shock period were qualitatively similar, but quantitatively less. In each case, dogs which failed to survive the experimental period were excluded from the results. Since the blood loss incurred by sampling was prohibitive, all determinations were not performed in every animal. The number of dogs studied for each parameter is shown in the column adjacent to the control values.

Table 8-II indicates the severity of the shock, with a statistically significant fall in blood pressure, cardiac output, and hematocrit and rise in pulse rate and respiratory rate. Deviations from control levels in oxy-

TABLE 8-II
CARDIOVASCULAR AND RESPIRATORY PARAMETERS
MEAN ± STANDARD DEVIATION
SECOND SHOCK PERIOD

	No. of Dogs	*Control*	*End Shock Period*
Blood pressure (mm Hg)	118	142.5 ± 24.1	$p < 0.001$ 32.8 ± 12.2
Pulse rate/min	53	167.4 ± 27.9	$p < 0.005$ 187.1 ± 41.9
Cardiac output (liters/min)	109	2.58 ± 0.89	$p < 0.001$ 0.60 ± 0.20
Respiratory rate/min	132	16.3 ± 12.4	$p < 0.001$ 37.4 ± 19.4
Hematocrit	126	39.7 ± 5.4	$p < 0.001$ 24.6 ± 4.9

gen tension and acid-base parameters are shown in Table 8-III. The oxygen tension increased in shock, concomitant with hyperventilation and a possible shift of the oxyhemoglobin dissociation curve to the right in the presence of a high H^+ ion concentration. Respiratory efforts failed to compensate completely for metabolic acidosis and the pH fell.

TABLE 8-III
ACID-BASE BALANCE
MEAN ± STANDARD DEVIATION
SECOND SHOCK PERIOD

	No. of Dogs	*Control*	*End Shock Period*
PaO_2 (mm Hg)	128	88.03 ± 11.59	$p < 0.001$ 103.03 ± 19.21
pH	128	7.32 ± 0.06	$p < 0.001$ 7.27 ± 0.14
$PaCO_2$ (mm Hg)	126	46.89 ± 7.41	$p < 0.001$ 25.66 ± 11.26
Total CO_2 (mEq/liter)	107	24.26 ± 3.28	$p < 0.001$ 11.36 ± 5.83

The changes in intermediary carbohydrate metabolism in its anaerobic and aerobic phases and associated energy mechanisms are shown in Tables 8-IV and 8-V. Hemorrhagic shock was marked by significant

TABLE 8-IV
PARAMETERS OF ANAEROBIC AND AEROBIC CHO METABOLISM
MEAN ± STANDARD DEVIATION
SECOND SHOCK PERIOD

	No. of Dogs	*Control*	*End of Shock Period*
Glucose (mg/100 ml)	117	105.55 ± 16.96	$p < 0.001$ 422.55 ± 206.89
Pyruvate (mEq/liter)	134	0.066 ± 0.036	$p < 0.001$ 0.186 ± 0.058
Lactate (mEq/liter)	134	1.310 ± 0.462	$p < 0.001$ 7.707 ± 3.490
Excess lactate (mEq/liter)	134		3.755 ± 3.276
Citrate (mEq/liter)	94	0.324 ± 0.120	$p < 0.001$ 0.743 ± 0.281
α-ketoglutarate (mEq/liter)	34	0.012 ± 0.002	$p < 0.001$ 0.028 ± 0.009
Malate (mEq/liter)	14	0.080 ± 0.031	$p < 0.001$ 0.131 ± 0.038

TABLE 8-V
ENERGY MECHANISMS
MEAN ± STANDARD DEVIATION
SECOND SHOCK PERIOD

	No. of Dogs	*Control*	*End of Shock Period*
Inorganic phosphate (mEq/liter)	114	2.830 ± 1.201	$p < 0.001$ 3.329 ± 1.241
AMP (mEq/liter)	112	0.022 ± 0.081	0.020 ± 0.049
ADP (mEq/liter)	112	0.066 ± 0.017	$p < 0.005 > 0.001$ 0.076 ± 0.035
ATP (mEq/liter)	116	0.340 ± 0.063	$p < 0.001$ 0.258 ± 0.067

increases in glucose, lactate, pyruvate, citrate, α-ketoglutarate, and malate. Associated with the elevated lactate and pyruvate levels were a statistically significant rise in inorganic phosphate and ADP values and a progressive fall in arterial ATP levels. Values for AMP showed little deviation from the controls.

Table 8-VI shows the values for NAD^+ and NADH for the control period and the end of the second shock period. There was a statistically significant (p. < 0.001) fall in NAD^+, with no appreciable change in NADH.

TABLE 8-VI
VALUES OF NAD^+ AND NADH
MEAN VALUE ± STANDARD DEVIATION
(mEq/liter)

	Control	*End Second Shock Period*
NAD	0.0711 ± 0.0244	$p < 0.001$ 0.0375 ± 0.0173
NADH	0.0118 ± 0.0084	0.0123 ± 0.0063

The increased levels of total and excess lactate indicate that a shift from aerobic to anaerobic carbohydrate metabolism does occur in severe hemorrhagic shock. Additional evidence for diminution of Krebs cycle activity is afforded by the significant rise in citrate, α-ketoglutarate, and malate levels. Although transposition of human and animal data is not always possible, the preoperative control values for the three acids are almost identical in the two species. Since Schweizer and Howland[7] showed that changes in the levels of the Krebs cycle acids do not occur in the anesthetized patient in the absence of shock, the elevated values in the present study suggest that hemorrhagic shock may interfere with the transfer of hydrogen ion from the substrate acids to the intramitochondrial electron transport chain.

Although the levels of total and excess lactate indicate the extent of conversion to anaerobic oxidation and also produce significant degrees of metabolic acidosis, the diminution in ATP and NAD^+ levels probably represent a more basic and important factor in the prognosis of hemorrhagic shock. In normal aerobic carbohydrate metabolism, 95 per cent of the ATP is formed within the mitochondria, coupled with the intramitochondrial electron transport chain. A shift to anaerobic metabolism would largely eliminate this source of ATP, leaving only the small amount of ATP associated with the glycolytic process. A significant fall in arterial ATP levels was noted in the shock dogs. Since ATP is the main storehouse of energy in the body, cellular survival in anaerobic states is dependent upon the metabolism of increased amounts of glucose by the series of enzymatically controlled reactions in the glycolytic chain.

A key reaction in this chain is the oxidation of glyceraldehyde 3-P to 1,3-diphosphoglycerate. In this reaction, catalyzed by glyceraldehyde 3-P dehydrogenase, oxidized NAD^+ acts as the hydrogen acceptor and is converted to NADH. Since reoxidation of NADH via the intramitochondrial electron transport chain cannot occur under anaerobic conditions, the formation of oxidized NAD^+ is dependent upon the extramitochondrial reduction of pyruvate to lactate.

A significant decrease in arterial levels of NAD^+ was noted in the shock dogs, but it is impossible to determine whether the decrease occurred in aerobic (Krebs cycle) or anaerobic (glycolytic) NAD^+. Since the B-hydroxybutyrate/acetoacetate ratio reflects the intramitochondrial $NADH/NAD^+$ ratio, knowledge of the values of these metabolites would be necessary to clarify the status of aerobic NAD^+.

Although intramitochondrial NAD^+ levels are of great importance in aerobic respiration, a reduction in anaerobic NAD^+ production would constitute a more serious threat to the organism in hemorrhagic shock with its dependence on glycolysis for ATP production. A decrease in anaerobic NAD^+ could be due to several factors including inhibition of the enzyme catalyst, lactic dehydrogenase, a decline in the supply of glucose or glycogen or failure of the body to mobilize reserve stores of glucose. Since diminution of anaerobic NAD^+ may play a role in the reversibility of hemorrhagic shock, the subject warrants additional investigation to clarify the present lack of definitive information.

REFERENCES

1. Schatz, G.: Mitochondrial oxidative phosphorylation. *Angew Chem* [*Engl*], *6*:1035, 1967.
2. Lehninger, A.L.: *The Mitochondrion* New York, Benjamin Co, 1965.
3. Howland, W.S.: Personal communication.
4. Elliott, K.A.C.: Intermediary metabolites and respiratory catalysis. *Physiol Rev*, *21*:267, 1941.

5. Bergmeyer, H.V.: *Methods of Enzymatic Analysis*. New York, Academic Press, 1963.
6. Huckabee, W.E.: Relationships of pyruvate and lactate during anaerobic metabolism: I: Effects of infusion of pyruvate or glucose and of hyperventilation. *J Clin Invest*, *37*:244, 1958.
7. Schweizer, O.; Howland, W.S.; Sullivan, C., and Vertes, E.: The effect of ether and halothane on blood levels of glucose, pyruvate, lactate and metabolites of the tricarboxylic acid cycle in normotensive patients during operation. *Anesthesiology*, *28*:814, 1967.

Chapter 9

Treatment of Anticholinesterase Poisoning

James S. Ketchum

There is little doubt that compounds capable of inhibiting cholinesterase enzymes have moved from a position of mere academic interest to one of considerable public health significance within the past two decades. The reasons are to be found not only in the expanding production and use of organophosphorus and carbamate agents, but from the widespread popularity, in nations such as Japan and Denmark in particular, of anticholinesterase poisons as implements of suicide. Incomplete statistics suggest that several thousand purposeful ingestions of lethal doses of pesticides have occurred throughout the world within the past few years. Yet, many physicians have a very limited understanding of the effects of these agents and feel quite inadequate with respect to their treatment. Today, therefore, I wish to summarize some current theories of the mode of action and current recommendations about the therapeutic management of anticholinesterase poisoning, with some emphasis on those aspects which may interest the clinician.

Several excellent discussions of the problem of anticholinesterase poisoning and its treatment have been published in recent years by Koelle,[1] Grob,[2] Ellin and Wills,[3] Sim,[4] Frost and Poulsen,[5] Juul and Spiers,[6] to name just a few. It would be futile to attempt to improve upon these detailed and authoritative expositions. I will therefore cover briefly the concepts with which there is general agreement, leaving the remainder of the time to discuss some interesting discrepancies and areas of ignorance.

PHYSIOLOGY AND PHARMACOLOGY

Of the numerous subspecies of cholinesterase found in the body, true cholinesterase (or acetylcholinesterase, to use the terminology proposed by Dixon and Webb[7]) is generally thought to be the most important in neural transmission. Besides being found in the red cell, where its function is obscure, it has been identified on both sides of the synapse throughout most of the peripheral and central divisions of the nervous system. While the question of the role of acetylcholine (ACh) at these synapses is far from settled, it seems safe to say that it plays an important part not only in transmission at the neuromuscular junction, in autonomic ganglia, and in parasympathetic postsynaptic nerve endings, but also in many central synapses ranging from the Renshaw interneurons of the spinal cord to the pyramidal cells of the cerebral cortex. The inhibition of acetylcholinesterase in any of these locations appears to facilitate the excitation of postsynaptic elements, presumably, but not necessarily, by an acetylcholine sparing action which allows the neurotransmitter to act more persistently upon the postsynaptic membrane.

Pseudocholinesterases, which are more efficient in hydrolyzing choline esters containing larger acyl groups (such as propionyl or butyryl), are found in plasma, in glial tissue, and in some nerve cells, but the function of these nonspecific cholinesterases remains a mystery.

Acetylcholinesterase is a protein with molecular weight of about 260,000, and a structure that is only partially known. Its role in the postsynaptic membrane is to bind acetylcholine molecules one at a time and cause them to be split by hydrolysis (Fig. 9–1). The active site at which this occurs contains an anionic portion which receives the choline end of the ACh molecule and an adjacent esteratic site where the acetyl moiety is bound. Choline first splits away and acetic acid is created a few microseconds later by a molecule of water which breaks the remaining bond at the esteratic site.

The mechanism of poisoning involves a competitive inhibition of this process. When an organophosphorus compound such as diisopropyl fluorophosphate (DFP) is introduced into the circulation, the enzyme attacks the electrophilic phosphorus atom, eliminating an acid group (Fig. 9–2). In this bimolecular nucleophilic substitution reaction, the phosphorus forms a covalent bond with the nucleophilic group on the enzyme. Instead of an acetylated, a phosphorylated enzyme is formed which is quite stable. Some agents, such as sarin, become still less susceptible to reactivation by losing an alkyl group, a process known as dealkylation or "aging." Once this happens, weeks or months may elapse before spontaneous hydrolysis occurs, restoring the active site. Thus, while all of the organophosphorus poisons are regarded as irreversible inhibitors, some are more irreversible than others. True cholinesterase is less rapidly activated than pseudocholinesterase, and that portion found in the red cell may be restored only by replacement of the cell itself. At the synapse, the time course of recovery is less well documented, but seems to parallel the pseudocholinesterase in the plasma, a phenomena which may have clinical significance.

Many of the clinical consequences (Table 9-I) of cholinesterase inhibition are predictable from the hypothesis of neurohumoral transmission, at least insofar as peripheral neural systems are concerned. At muscarinic sites, ACh buildup stimulates the secretory and contractile

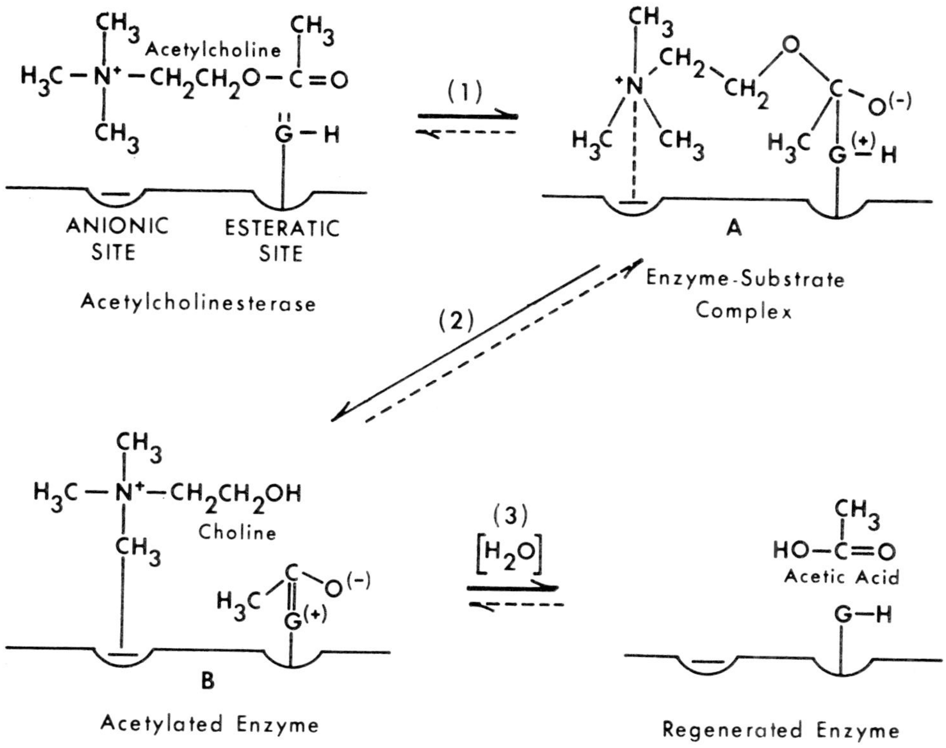

Figure 9–1. Steps involved in the hydrolysis of ACh by cholinesterase. The substrate, ACh, combines with an active unit of the enzyme (*step 1*) to form a complex (*A*) by electrostatic attraction between the N^+ atom of the choline moiety and the anionic site of the enzyme, and the electrophilic C atom of the carboxyl group and a protonated acidic group (—G—H) of the esteratic site; choline is then split off (*step 2*), leaving the acetylated enzyme (*B*), which reacts rapidly with water (*step 3*) to produce acetic acid and the regenerated active enzyme.
Heavy, light, and dash arrows represent extremely rapid, intermediate, and extremely slow or insignificant reaction velocities, respectively. (Reprinted from *The Pharmacological Basis of Therapeutics,* Third Edition, Eds. L.S. Goodman and A. Gilman, 1965 with permission of the publisher: The Macmillan Company, New York.)

activity of glands and smooth muscle fiber respectively. In the sympathetic ganglia, including the adrenal medulla, the rising levels of ACh cause sympathetic stimulation followed by depression. As the concentration of ACh increases, the neuromuscular junction displays a progression of nicotinic effects, beginning with heightened irritability and fasciculations, and proceeding to paralysis.

In the CNS the state of affairs is more complicated. Although considerable evidence has now been published suggesting the presence of cholinergic systems, both muscarinic and nicotinic, within various brain structures, the end result of cholinergic excess in these systems cannot

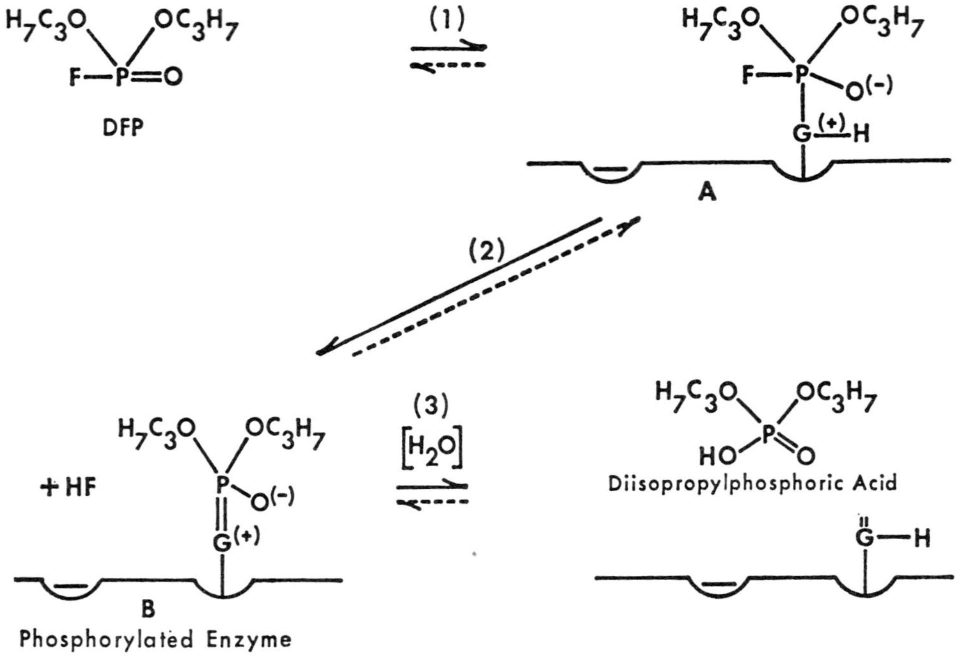

Figure 9–2. Steps involved in the inhibition of cholinesterase by organophosphorus agents. DFP and similar organophosphorus inhibitors react only at the esteratic site to form a phosphorylated enzyme; in the case of diisopropylphosphoryl-cholinesterase (*B*), essentially no spontaneous hydrolytic reactivation occurs.
Heavy, light, and dash arrows represent extremely rapid, intermediate, and extremely slow or insignificant reaction velocities, respectively. (Reprinted from *The Pharmacological Basis of Therapeutics,* Third Edition, Eds. L.S. Goodman and A. Gilman, 1965, with permission of the publisher: The Macmillan Company, New York.)

TABLE 9–I
SIGNS AND SYMPTOMS OF ANTICHOLINESTERASE POISONING

EYE	CENTRAL NERVOUS SYSTEM
Miosis	Headache
Lacrimation	Dreams
Dimness of vision	Fatigability and weakness
Impaired accommodation	Mood changes
Pain on accommodation	Increased perspiration
Injection	Dizziness
	Fasciculation
NOSE	Paresthesias
Rhinorrhea	Convulsions
RESPIRATORY SYSTEM	GASTROINTESTINAL TRACT
Constriction of chest	Anorexia and nausea
Cough	Vomiting
Wheezing	Diarrhea
Dyspnea	
GENITOURINARY TRACT	
Backache, frequency, or nocturia	

always be predicted, because of our limited knowledge of the details of nervous system organization. Clinically, dizziness, anxiety, nausea, nightmares, disturbances of concentration and memory, and depression of mood are typical behavioral manifestations of moderate poisoning. Disorientation, unconsciousness, hyperthermia, and respiratory arrest are observed after lethal doses.

The time course varies. Obviously the route of administration is one important factor. Lipid soluble agents may enter through the skin, causing the first symptoms many hours after contact, or sooner after large doses. The rate of absorption also governs the rate at which symptoms progress. While death in less than five minutes has occurred, often more than a day elapses before the full severity of the illness becomes evident. In the case of parathion (Fig. 9–3), itself relatively nontoxic, con-

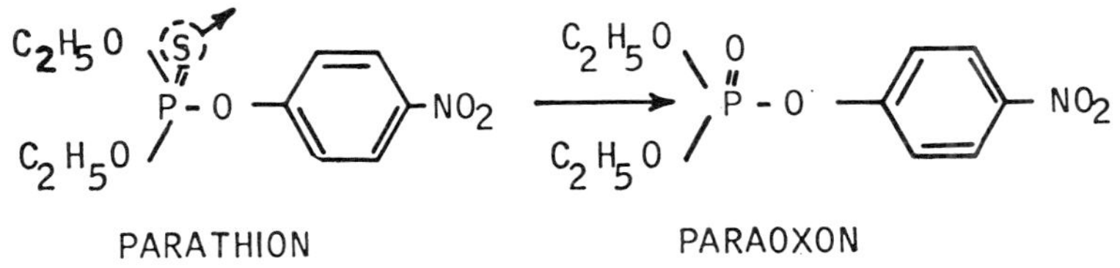

Figure 9–3. Conversion of parathion to paraoxon

version by liver enzymes to highly toxic paraoxon precedes the appearance of symptoms. Thus, in spite of their inexorable lethality, the action of anticholinesterase poisons is rarely so swift as to preclude effective medical intervention.

DIAGNOSIS AND TREATMENT

Diagnosis is not difficult, provided the physician is alert and takes an inquisitive history. In addition to the general signs of parasympathetic overactivity, local signs reflecting the route of exposure are helpful clues. Vapors, such as Sarin, will commonly cause rhinorrhea, constriction of the chest, and bronchospasm along with miosis, as the earliest symptoms. Liquid agents absorbed from circumscribed skin areas will initially produce local sweating and fasciculations. Ingested materials are likely to first cause cramps, diarrhea, vomiting, and hypermotility.

Mental changes are usually obvious in severe cases, and indeed may be the most reliable indicators of chronic low-grade exposure. Intellectual dullness, depression, memory impairment, troubled sleep, fatigability, and even hallucinations and delusions are features which should bring to mind the possibility of chronic poisoning. Cholinesterase levels may help resolve doubtful cases, particularly the red blood cell cholinesterase, which is less likely to be confounded by unrelated metabolic disturbances such as liver disease.

Treatment of anticholinesterase poisoning continues to be founded upon two cardinal principles: maintenance of adequate oxygen exchange and atropinization. To this nucleus may be added the use of oximes, a group of compounds which were found by Wilson and Ginsberg,[8] Davies and Green,[9] Askew,[10] and others to hasten the rate of removal of the organophorphorus inhibitor from its point of attachment to the esteratic site. Pyridine-2-aldoxime methochloride, better known as pralidoxime or 2-PAMCl, is one such compound. With one end anchored to the unoccupied anionic site, the free end of the 2-PAMCl molecule is brought close to a projecting side chain of the organophorphorus molecule (Fig. 9–4). The resulting forces of attraction weaken the inhibitor's hold on the esteratic site, facilitating its separation. Unfortunately this process does not work equally well for all poisons. It tends to work even less well when treatment is delayed, owing to the spontaneous tendency of some agents to undergo "aging," i.e. the loss of an alkyl group. But for many organophorphorus compounds, significant reactivation can be achieved by early treatment with intravenous or intramuscular 2-PAMCl.

Slow intravenous or intramuscular atropine should be given as soon as possible, in a dose of at least 2 mg, and repeated every five to ten

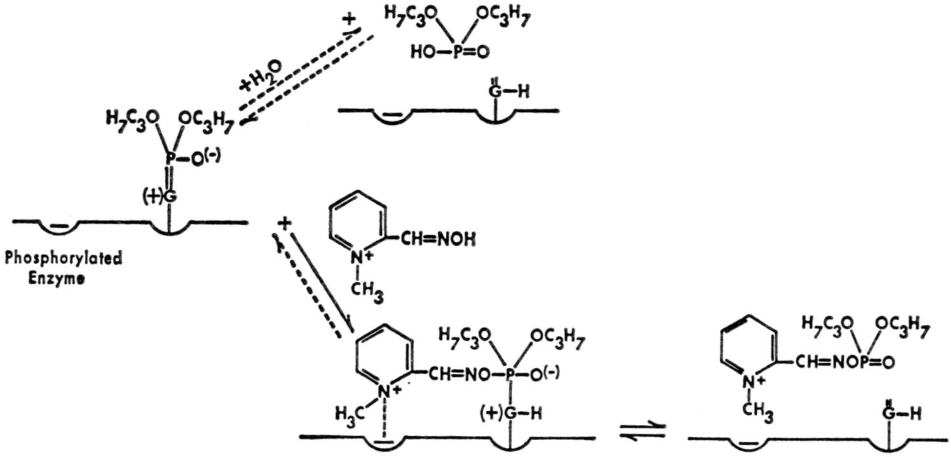

Figure 9–4. Chemical pathways for the reactivation of DFP-inactivated cholinesterase by pralidoxime (*below*) and by spontaneous hydrolysis (*above*). The rate of the latter reaction is insignificant. Symbols are same as used in Figure 9–1. (Reprinted from *The Pharmacological Basis of Therapeutics,* Third Edition, Eds. L.S. Goodman and A. Gilman, 1965, with permission of the publisher: The Macmillan Company, New York.)

minutes until definite atropine effects are evident. Most authors advocate the continuation of antimuscarinic treatment for at least twenty-four hours, which may require up to 50 mg or more in severe poisoning. On the other hand, some animal studies indicate that little is to be gained from overly massive doses, and indeed they may be antitherapeutic for reasons that are not well understood. Fear of atropine toxicity, however, should not lead to timidity in its use, since rarely does death occur at doses below 100 mg, and atropine delirium can be reversed readily by 4 mg of intramuscular physostigmine should inadvertent overdosage occur.

2-PAMCl is best administered intravenously as a 5% solution, in water, allowing up to thirty minutes for 50 cc of this concentration to run in. In a crisis, this dose of 2.5 gm can be given more rapidly, but not faster than 0.5 gm/min. Plasma levels above 20μg/ml are typical at thirty minutes following such dose, after which the rate of disappearance from the plasma is gradual. The half-life following oral doses is about 2.5 hours in normal volunteers according to Sidell[11] (Fig. 9–5). Estimates of the half-life after intravenous administration vary according to the portion of the time scale examined. Sidell also found[12] that 60 to 80 per cent of inhibited cholinesterase could be reactivated by as little as 0.25 gm of 2-PAMCl given twenty-four hours after poisoning and only slightly better results were achieved by giving much larger doses, or much earlier treatment (Fig. 9–6). Since the volunteers in this experiment received doses of the agent barely sufficient to produce mild symptoms (cholinesterase levels between 10 and 50 per cent of baseline), it cannot be concluded fairly that small doses of 2-PAMCl would be equally effective in more severe poisoning. What this study does suggest is that the observed benefits of larger doses of oxime may be related to factors other than the degree of reactivation achieved.

Most of the general measures advocated for the management of anticholinesterase poisoning are those which medical judgment alone might dictate. Removal of residual agent from the skin, stomach, clothing, and environment should be accomplished concomitantly with the initiation of specific therapy. Vigilance should be the watchword for at least twenty-four hours because of the possibility of delayed absorption, as well as sudden deterioration in patients who seem to be out of danger. Convulsions, gastric dilatation, acidosis, pneumonitis, and insomnia with psychic discomfort are complications which should be anticipated and treated in the usual manner. One caveat is that phenothiazine tranquilizers and barbiturates have supra-additive effects in combination with some cholinesterase inhibitiors.[13] Trimethadione is recommended for the control of seizures, and small doses of chloral hydrate can probably be used safely for sedation.

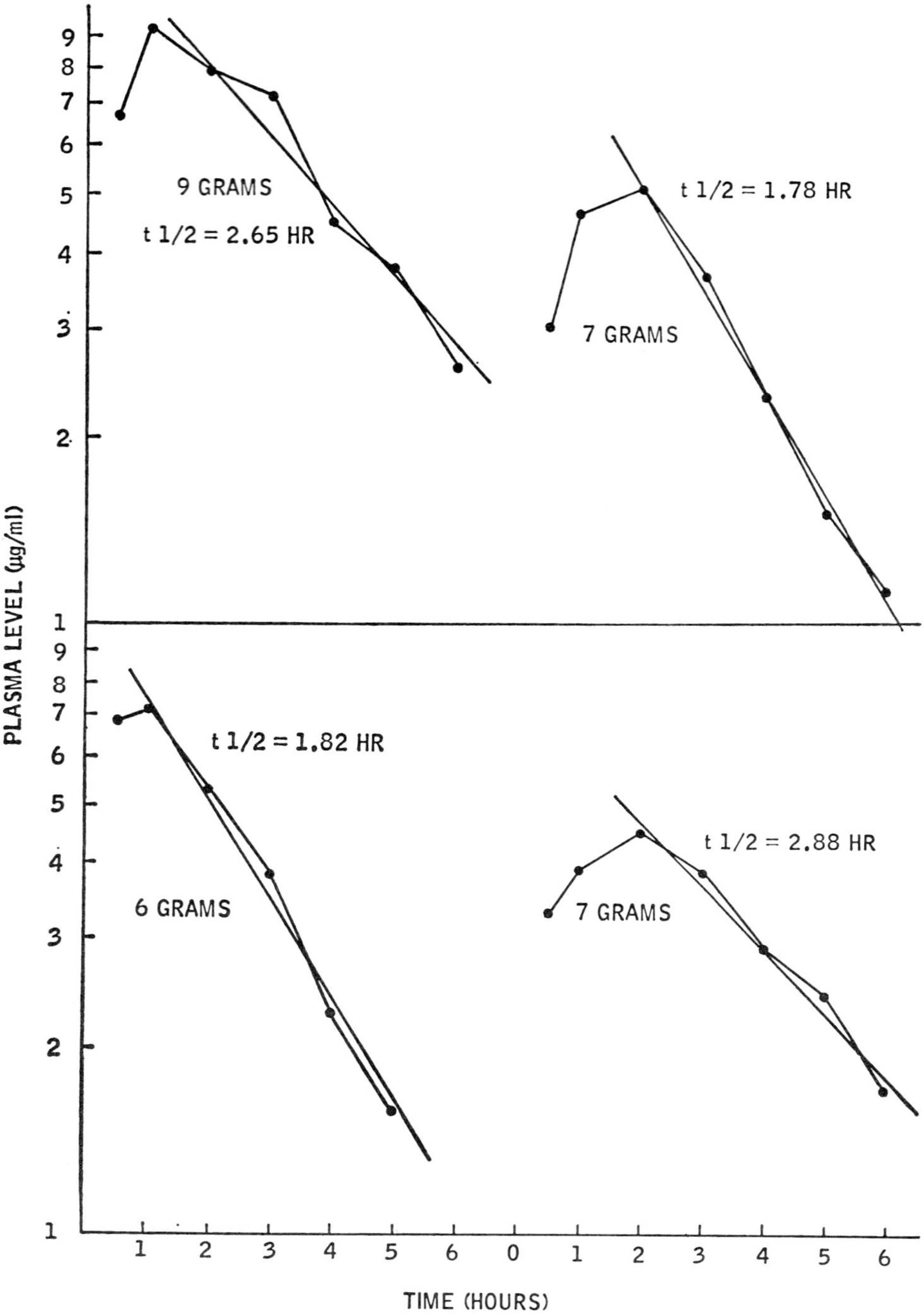

Figure 9–5. Plasma levels of 2-PAMCl after oral administration (Reproduced with permission of copyright owner. See reference 11.)

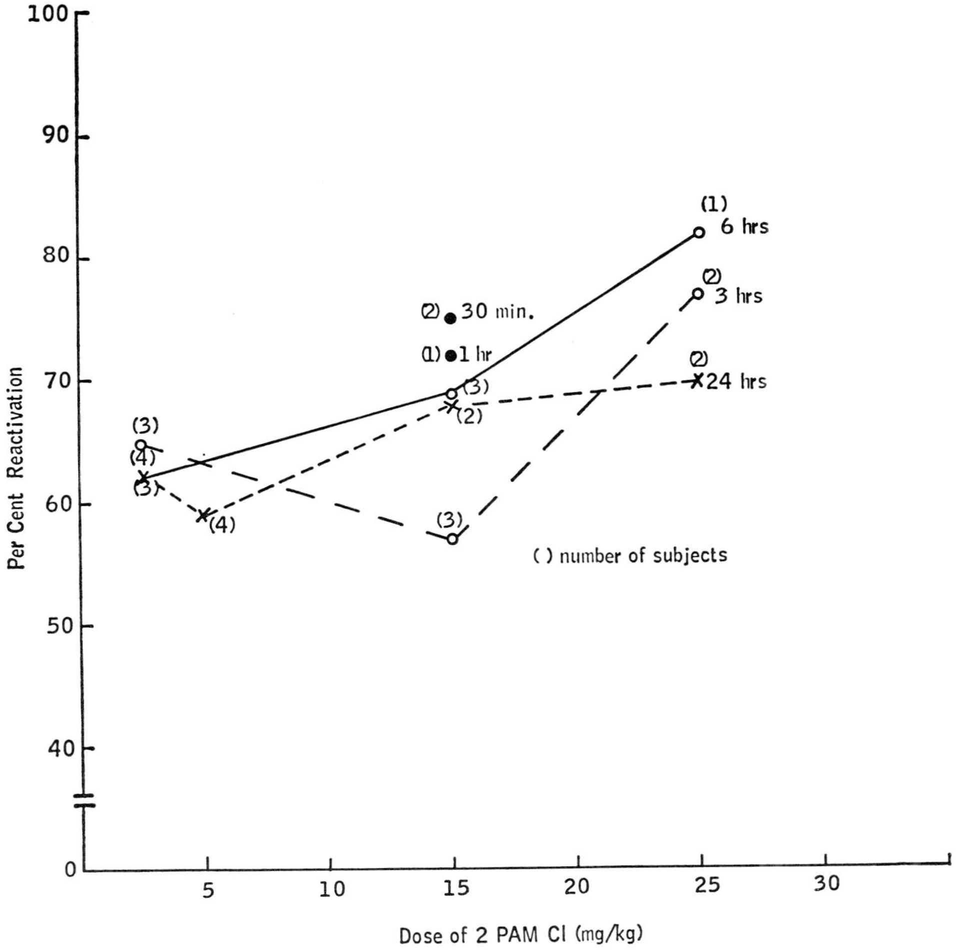

Figure 9–6. Reactivation of RBC-cholinesterase by intravenous 2-PAMCl after inhibition by intravenous VX. Time is hours after VX; percent reactivation is within one hour after oxime.

UNRESOLVED QUESTIONS AND CONTROVERSIES

In spite of much research and considerable progress, many aspects of therapy continue to provoke dispute, while a few seem to have been insufficiently disputed. One of these is the apparent discrepancy between theory and practice in the therapy of anticholinesterase agents. While almost every author points out that the acetylcholine-sparing effects of these poisons have nicotinic as well as muscarinic consequences and cause central as well as peripheral toxicity, pharmacotherapy continues to be based largely on atropine, which has insignificant nicotinic blocking action except at very high doses. Atropine, furthermore, has limited central potency compared to other muscarinic blocking agents such as scopolamine. Although it is often stated that oximes can deal with the nicotinic effects by reactivating the inhibited enzyme at the neuromuscular junction, we now realize that oximes actually have little or no ability to reactivate cholinesterase inhibited by agents such as soman, and are also unable to reverse the inhibition produced by certain carbamates. Furthermore, 2-PAMCl at therapeutic doses does not enter the CNS to any appreciable extent, and does not appear to reactivate brain cholinesterase, even when the inhibitor is otherwise oxime-responsive.

Yet even against oxime-resistant agents such as Soman, the addition of 2-PAMCl seems to be helpful. Bay[14] has suggested that this may be the result of 2-PAMCl's cocaine-like ability to block the re-uptake of liberated catecholamines, producing a salutory effect upon respiration and blood pressure and perhaps also hastening the distribution of atropine to its intended targets. If this is indeed how 2-PAMCl works in such situations, we are left in the uncomfortable position of advocating a treatment which gives the right result for the wrong reasons.

The choice of oxime is likewise a matter upon which no real consensus has been attained. 2-PAMCl has been endorsed primarily because it is considered less toxic and just as effective as other oxime preparations. But most of the toxicity comparisons have been based either on chronic administration to normal volunteers, which generally leads to gastrointestinal complaints within a few days, or on comparisons of the ratio between lethal and therapeutic doses in animals.

Since treatment with oximes is in practice not continued for more than a few hours, the chronic effects seem somewhat irrelevant. Similarly, the safety margin, provided it is adequate, seems unimportant in a situation where dosage is controllable. Opiates, muscle relaxants, and other drugs may be used safely with appropriate caution in doses which permit a much smaller margin for error.

This seems worth emphasizing because much published evidence suggests that other oximes, such as diacetylmonoxime (DAM), 1,1′-trimethylene-bis-(4-formyl pyridinium bromide)-dioxime (TMB-4), and bis[4-(hydroxyiminomethyl)pyridinium-(1)-methyl] ether, dichloride (Lu-H6), may be more efficacious than 2-PAMCl against particular organophosphorus and carbamate cholinesterase inhibitors. There is some indication, furthermore, that some of these other oximes may enter the nervous system more readily than 2-PAMCl. Thus it would appear more rational to design therapy to combat individual agents rather than to seek a single oxime for the treatment of all agents. This approach has certainly paid dividends in the field of antibiotic, antiepileptic, and antipsychotic drugs to mention only the more obvious examples.

To extend the discussion on this point a bit further, it appears in some of the published work on the comparative efficacy of oximes that important variables are often ignored. For example, in comparing TMB-4 or Lu-H6 with 2-PAMCl, little attention seems to have been paid to the fact that blood levels of oxime achieved by equal doses of these three agents, even by the intravenous route, are distinctly different. Plasma concentrations produced by 2-PAMCl in a series of studies by Calesnick *et al.*[15] were less than one-third those produced by TMB-4. Bay[16] found that 2-PAMCl given to monkeys produced plasma levels about half those produced by equal doses of Lu-H6. These differences, which are not solely attributable to differences in molecular weight or number of oxime groups per molecule, suggest that TMB-4 and Lu-H6 may be pharmacokinetically distinct from 2-PAMCl.

Similarly, it may not be fair to compare 2-PAMCl and the bis-quaternary oximes at equal doses, because in spite of approximately equal numbers of oxime groups per milligram of material, the bis-quaternary oximes can make use of only one of the two groups on each molecule at a time in the process of reactivation. Studies by Fleisher *et al.*[17] in which compensation in dosage was made on the basis of this reasoning have sometimes shown a superiority for Lu-H6.

On the other hand, Bay's results indicate that Lu-H6 is more rapidly eliminated from the circulation than is 2-PAMCl. This should be compensated for in the design of comparative studies by assuring the maintenance of parallel concentrations of the two agents throughout the period of observation.

Finally, although several investigators have concluded that the rate of reactivation is only slightly influenced by oxime concentration, above a certain minimum, it has also been shown by Fleisher *et al.*,[18] O'Leary *et al.*,[19] and others that larger doses of oximes, whether 2-PAMCl, TMB-4, or Lu-H6, do indeed provide protection against significantly larger multiples of the LD_{50}. This reinforces further the idea that therapeutic effect is not simply a matter of enzyme reactivation.

DIRECTIONS FOR FUTURE RESEARCH

Thus, while progress has been made in the management of anticholinesterase poisoning, the situation is far from ideal. This is not difficult to understand when one considers that treatment of enzyme inhibition is tantamount to treatment of the entire nervous system, where a single theme may appear in infinite variation. There are, nevertheless, certain aspects of the problem which would seem to deserve further inquiry.

One of these is the evaluation of various nicotinic blocking agents as further adjuncts to atropine and oxime therapy. Bay has recently shown in monkeys that central respiratory depression induced by intracisternal application of small doses of Soman is more effectively reversed by a combination of mecamylamine and atropine than by atropine alone. The omission of antinicotinic agents from the current regimen is a curious one.

Similarly, the possibility of employing agents other than atropine—scopolamine, for instance—should be explored further. Scopolamine is roughly equal to atropine in peripheral antimuscarinic potency, but eight times as potent centrally.[20] It would seem for this reason to offer certain advantages. Recently, a thirty-two-year-old man[21] sustained an accidental exposure to soman. Although in critical condition at the time of his admission, he responded to intravenous atropine, and by the second day appeared to be out of danger, although his red cell cholinesterase remained at zero. He continued to be dizzy, apprehensive, and slept and ate poorly. Mentally he was sluggish and performed calculations poorly. When given small doses of scopolamine by injection, his mental status improved noticeably and his performance on standard arithmetic tests rose significantly (Fig. 9-7). Treatment with similar doses of methscopolamine failed to produce the same benefit.

Apart from the apparent superiority of scopolamine over atropine in reversing the mental impairment and dysphoria associated with anticholinesterase poisoning, this case brings two other points to our attention. First, as is often the case, this individual showed great clinical improvement in the absence of measurable reactivation of cholinesterase. Perhaps this is somehow related to the phenomenon often observed in individuals who suffer frequent mild exposures to organophosphorus poisons. These workers are sometimes found to have cholinesterase levels approaching zero, yet only mild symptoms. Presumably some sort of adjustment is made by the nervous system, such as the mobilization of reserve cholinesterase, but this is a speculation which deserves more intensive study. Second, and more germane to clinical management, there seems to have been relatively little research directed toward the reduction of morbidity in patients who have survived the acute episode of poisoning. Perhaps it is natural to feel grateful that a fatal outcome has been averted. But many weeks of disability and discomfort may precede the return to work because of lingering autonomic and behavioral symptoms. This is costly to both patient and employer, and there are indications, illustrated in this case, that measures can be taken to reduce this incapacitation. Surely still more can be achieved.

While these suggestions do not exhaust the list of worthwhile avenues

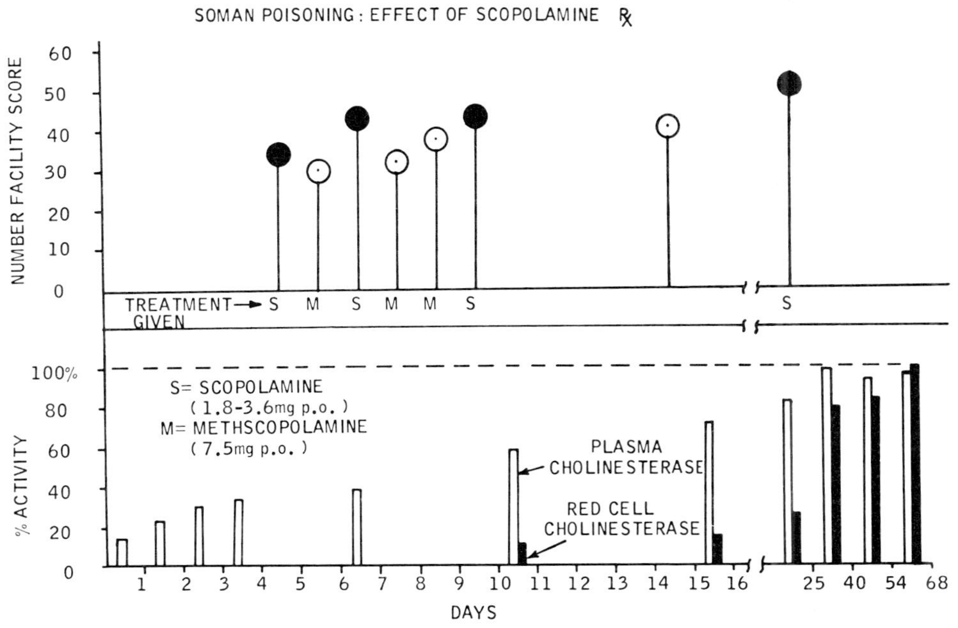

Figure 9–7. A summary of the course of a patient accidentally exposed to Soman. Note that on days when scopolamine treatment was given, the average Number Facility scores were higher (*black dots*) than on days when methscopolamine was substituted (*open circles*).

to be explored in the area of therapy for the anticholinesterase agents, they may suffice to indicate that work in this interesting and vital field is far from finished.

REFERENCES

1. Koelle, G.B. (sub-editor): Handbuch der experimentellen Pharmacologie. No. 15. Cholinesterases and anticholinesterase agents. Springer Verlag, Berlin. 1963.
2. Grob, D.: The manifestations and treatment of poisoning due to nerve gas and other organic phosphate anticholinesterase compounds. *Arch Intern Med, 98:* 221–239, 1956.
3. Ellin, R., and Wills, H.: Oximes antagonistic to inhibitors of cholinesterase. *J Pharm Sci, 53:*995–1007 and 1143–1150, 1964.
4. Sim, V.M.: Chemicals used as weapons in war. In DiPalma, J. (Ed.): *Drill's Pharmacology in Medicine.* New York, McGraw-Hill, 1965.
5. Frost, J., and Poulsen, E.: Poisoning due to parathion and other organophosphorus insecticides in Denmark. *Dan Med Bull, 11:*169–177, 1964.
6. Juul, P., and Spiers, F.: Cholinesterase inhibitors: pharmacology, symptoms of poisoning, and treatment. Studies on the cholinesterase activities in glaucoma patients. *Milit Med, 132:*501–511, 1967.
7. Dixon, M., and Webb, E.C.: *Enzymes.* New York, Academic Press, 1958.
8. Wilson, B., and Ginsberg, S.: Powerful reactivator of alkylphosphate-inhibited acetylcholinesterase. *Biochim Biophys Acta, 18:*168, 1955.

9. Davies, D., and Green, A.: The kinetics of reactivation, by oximes, of cholinesterase inhibited by organophosphorus compounds. *Biochem J, 63*:529–535, 1956.
10. Askew, B.: Oximes and atropine in sarin poisoning. *Br J Pharmacol, 12*:340–343, 1957.
11. Sidell, F.R.; Groff, W.A., and Ellin, R.I.: Blood levels of oxime and symptoms in humans after single and multiple doses of 2-PAMCl. *J Pharm Sci, 58*:1093, 1969.
12. Sidell, F.R.: Edgewood Arsenal, Md. Unpublished data.
13. Arterberry, J.D.; Bonifaci, R.W.; Nash, E.W., and Quinby, G.E.: Potentiation of phosphorus insecticides by phenothiazine derivatives. *JAMA, 182*:848–850, 1962.
14. Bay, E.: The action of 2–PAMCl on catecholamine-induced responses. *Pharmacologist, 4*:149, 1962.
15. Calesnick, B.; Christensen, J.A., and Richter, M.: Human toxicity of various oximes: 2–pyridine aldoxime methyl chloride, its methane sulfonate salt and 1,1-trimethylenebis-(4-formylpyridinium chloride). *Arch Environ Health, 15*:599–608, 1967.
16. Bay, E.: Edgewood Arsenal, Md. Unpublished data.
17. Fleisher, J.H.; Harris, L.W.; Miller, G.R.; Thomas, N., and Cliff, W.J.: Antagonism of Sarin poisoning in rats and guinea pigs by atropine, oximes, and mecamylamine. (Accepted for publication in *Toxicol Appl Pharmacol.*)
18. Fleisher, J.H.; Michel, H.O.; Yates, L., and Harrison, C.S.: 1,1′–Trimethylene bis (4-formylpyridinium bromide) dioxime (TMB4) and 2-pyridine aldoxime methiodide (2-PAM) as adjuvants to atropine in the treatment of anticholinesterase poisoning. *J Pharmacol Exp Ther, 129*:31–35, 1960.
19. O'Leary, J.F.; Kunkel, A.M., and Jones, A.H.: Efficacy and limitation of oxime-atropine treatment of organophosphorus anticholinesterase poisoning. *J Pharmacol Exp Ther, 132*:50–57, 1961.
20. Crowell, E.B., and Ketchum, J.S.: The treatment of scopolamine-induced delirium with physostigmine. *Clin Pharmacol Ther, 8*:409–414, 1967.
21. Sidell, F.R.: Edgewood Arsenal, Md. Unpublished case report. February 1969.

Chapter 10

Suppression of Adrenal and Pituitary ACTH Function by Administration of Corticosteroids

Don H. Nelson

Since corticosteroids first came into use some twenty years ago there has been a growing employment of them for a wide number of conditions. In fact, I have seen statistics which indicate that they may be the most widely used pharmaceutical products in the United States. A large number of patients receive one type or another of corticosteroid without difficulty, but there are enough problems in their use that we have to consider them.

Here, we are going to pay attention to those patients who, following administration of corticosteroids, develop adrenal atrophy. Of course, all patients receiving corticosteroids do this, but there is a small group of patients who require special attention from the anesthesiologist and the surgeon. These are they who, following administration of corticosteroids, fail to have rehypertrophy of their glands and who come to surgery with adrenal insufficiency of a mild type until exposed to stress and then develop a picture of addisonian crisis. A few points are important to keep in mind in this condition. These patients do not have the typical appearance of Addison's disease. They do not become pigmented. They usually do not give the symptomatology of gastrointestinal disturbance, of severe weakness, or malaise prior to the time that they are recognized. These patients all produce some corticosteroids—enough so that they do not show these obvious signs of Addison's disease but too little to maintain them during periods of stress. If one measures urinary corticosteroids in this group of patients, one finds that they have low-normal values but not values that are clearly abnormal. The only way that one can definitely make the diagnosis is by carrying out an ACTH stimulation test as we do in the diagnosis of Addison's disease. This consists of giving intravenous ACTH and measuring either plasma or urinary corticosteroids. This is something that cannot be undertaken in every patient who receives corticosteroids and comes for a surgical procedure. When these patients are subjected to stress, be it in the operating room or due to some infection or injury, they develop the typical picture of an Addisonian crisis—their blood pressure falls and they often develop fever and go into shock. It is at this time that the surgeon, the anesthesiologist, and the physician caring for these individuals are faced with the question, Do they have adrenal atrophy; should they have corticosteroids? If the decision is made to give them corticosteroids, the obvious method of treatment is intravenous hydrocortisone in a dose of 100 to 300 mg given intravenously. My own preference would be to give 100 mg Stat I. V. and then start a drip with 200 or 300 mg in it. Other supportive therapy that is indicated for shock should also be given; and if the patients have infection, this should be treated appropriately; but ACTH should not be given, as it will be ineffective.

Chronically in an attempt to prevent adrenal atrophy, we reduce the corticosteroids gradually, although there is no good evidence at the moment which proves that this prevents long-term atrophy. Some individuals have suggested giving ACTH to these patients, but again there is no evidence that this prevents the long-term atrophy. Certainly any patient with atrophic adrenals given ACTH will have hypertrophy of these glands; but when the ACTH is discontinued, the likelihood is that their adrenals will again atrophy.

Experimentally, it has been demonstrated that intravenous ACTH is not useful in this type of patient. When the gland is deprived of ACTH either following hypophysectomy or corticosteroid therapy, it becomes atrophic and does not respond acutely to ACTH administration. With chronic administration over a period of days, it will hypertrophy, however, and then respond to ACTH. So, if one is faced with a patient with adrenal atrophy and one gives ACTH acutely, no secretion will occur. If one gives it chronically, one can rehypertrophy the gland and secretion will return.

How much corticosteroid suppresses adrenal function? We know that the normal adrenal gland in an adult produces approximately 20 mg of cortisol per day. This is directly under the control of ACTH; and thus, if one gives enough cortisol or other corticoid to suppress ACTH, essentially that which is equivalent to the amount normally produced, then one will no longer produce ACTH and adrenal atrophy will occur. Somewhere in the neighborhood of 30 mg of cortisol equivalent given continuously through the twenty-four hour period replaces the normal output of the adrenal gland, suppresses ACTH, and results in adrenal atrophy. This is not a very large amount of corticosteroid. This is less than 10 mg of prednisone for instance and only about 1.5 mg of dexamethasone.

The question is often asked, How long does it take to produce adrenal atrophy? Adrenal atrophy can be produced in animals in a matter of two or three days. We do not have good statistics on patients but there is no reason to suspect that it takes longer. Different corticosteroids have different lengths of action. Cortisone effects last only about twelve hours.

Some of the substituted compounds last a little longer, perhaps up to eighteen hours. If one gives I.M. cortisol acetate, the effect may last up to four weeks. There has developed some enthusiasm for the use of intermittent corticosteroids with the idea that thus one does not maintain suppression of ACTH continuously, that the adrenal gland will not become so atrophic, and that it will be more responsive. Indeed, there is evidence that this is so and thus treatment of patients with corticosteroid should be carried out with the smallest dose of hormone at the longest interval that will do the job. Oftentimes, however, that is not a very small dose and it is not a very long interval.

One can assume that it will take somewhere between one and three weeks following discontinuance of corticoid therapy before the normal subject will have adrenals which will respond as they did before administration of the corticosteroids. The small percentage of patients who fail to respond and continue to have atrophic adrenal glands are of the greatest concern, and the approach to them is the important question to be answered. They have adrenal atrophy, but what is the etiology of this suppression? Is the primary difficulty in the pituitary gland or is it in the hypothalamic centers which stimulate the pituitary gland to secrete ACTH? The pituitary gland can secrete ACTH in response to stress in these patients, so the primary defect probably is not in the pituitary gland. The fact that the adrenal gland secretes corticosteroid very well after enough ACTH is given over a period of days suggests that we have not produced a primary defect in the adrenal glands either, and thus it is highly probable that the primary defect is in the hypothalamus.

Normally, the adrenal gland responds to diurnal variation in the secretion of ACTH. The highest levels of ACTH occur early in the morning with the output of most of the corticosteroids during the early part of the day. The levels then fall off toward evening and during the night again begin to rise with production of increased quantities of ACTH and increased corticosteroid secretion. In some patients who have received corticosteroids for a long period of time, this normal diurnal secretion of ACTH is interfered with. The very large doses of corticosteroid which have been given have dampened this control center so that it is no longer sensitive to the low levels of corticosteroid which are present and therefore does not signal the pituitary gland to put out ACTH in the normal diurnal rhythm. Because the pituitary does not do this, the adrenal glands become atrophic. As I pointed out before, however, the adrenal gland continues to secrete small quantities of corticosteroid, probably due to some very low rate of secretion of ACTH. Very importantly, aldosterone, which is one of the major secretory products of the adrenal cortex, is not chiefly under control of the pituitary gland and ACTH, but rather is controlled by the renin angiotensin mechanism. Thus, normal quantities of this salt-retaining hormone are produced in these subjects and they do not develop the electrolyte disturbances that we associate with other types of adrenal insufficiency.

The surgical management of the patients who have recently received corticosteroids is important. If patients are currently receiving steroids, there is no question that the dosage should be increased. These patients have adrenal atrophy and are not going to respond to the stress with increased corticoid secretion. They will become adrenally insufficient if corticosteroids are not given in increased amounts. These should be given intravenously during the operation and perhaps intramuscularly the night before. In most cases, they can be returned to their maintenance dose over a period of three or four days postoperatively. Our own approach is usually to give 300 mg during a major operation, cut that to 150 the day after, 75 the day after that, and 37.5 or whatever their maintenance dose is on the following day. This works out very well for the patient who is currently receiving corticosteroids and who, if he does not receive increased corticosteroid during the operation, will become adrenally insufficient.

If the patient has not received corticosteroids for a month or more, it can be assumed that the adrenals will have returned to normal function in most cases. In one large institution that does a lot of consultative surgery, the policy has been established that everyone who has ever received an adrenal suppressive dose of corticosteroids receives corticosteroids during operation. I would take a little exception to this policy because of the very small percentage of patients that need corticosteroids. There are complications in giving large doses of corticosteroids for a few days which may include bleeding, psychological disturbances, and infection. So rather than give corticosteroids to all such patients, we prefer to have corticosteroids available for any patient who has received adrenal suppressive doses of corticosteroids at any time in the past. If they are given early to a patient with adrenal insufficiency who is developing hypotension, the response is dramatic. This is usually the case if the condition is recognized and corticosteroids are given intravenously in good dosage.

May I conclude by pointing out that there is nothing that takes the place of a good history. If prior to receiving anesthesia, the patient, the physician, and the surgeon are always asked about corticosteroids, the anesthetist can be prepared and difficulties with adrenal insufficiency will be few.

Chapter 11

Hazards of Steroid Associated with Anesthesia

Tsutomu Oyama

The report in 1952, by Fraser *et al.* on the postoperative collapse and death of a thirty-four-year-old man treated for rheumatoid arthritis with corticosteroid therapy for eight months prior to hip arthroplasty was soon followed by many others.[1-11] Since the latter half of the last decade our anesthetic colleagues have been told about the dangers of adrenal collapse. A myth has developed regarding steroid cover, i.e. all patients who have received steroid therapy should have steroid cover.

The present study was undertaken to investigate two questions: (1) To what extent and how often do abnormalities of adrenal cortical function occur during the stress of anesthesia and surgery in steroid-treated patients? and (2) To what extent can these abnormalities be correlated with the observed blood pressure response during surgery.

Reviewing forty-three reported adrenal crises during or after operation, I found the following points:

1. Most of the reported cases occurred in the early period between 1952 and 1958 when the intermittent or small amount of maintenant administration of cortisone was not established.
2. Only two patients received less than 1 gm of total steroid equivalent to cortisol, and the remaining patients (75%) received more than 1 gm.
3. There were only six patients (14%) who received less than 26 mg of cortisol. This would suggest that the patient whose daily dose of cortisol was less than 26 mg had less episode of adrenal failure than those whose doses were more than 26 mg.
4. The duration of steroid treatment for less than one month was 32 per cent, and less than six months was 62 per cent of total reported cases.
5. Twenty-six patients (60%) had been treated with steroid either continuously or within one week prior to operation, but none of them received steriod on the day of operation.

Now I should like to present our data. Thirty-nine surgical patients were studied. Nineteen patients were currently receiving corticosteroid therapy and the remaining twenty patients who had never received corticosteroid served as controls. The nineteen patients who were receiving corticosteroid had this therapy in varying dosage for periods ranging from one week to five years (mean 6 months). The estimated mean daily equivalent dosage of cortisol was 80 mg (range 30–300 mg). The amount of preoperative steroids, the time between withdrawal of these drugs and operation, and the type of operation are shown in the Table 11-I. Corticosteroid therapy was discontinued twenty-four hours prior to surgery in all patients.

The anesthesia was started at 8:30 A.M. to minimize the effect of the normal diurnal variation of plasma cortisol. The anesthesia was induced and maintained with halothane (0.5%–2%). N_2O (2 l/m) and O_2 (2 l/m), combined with intermittent injection of SCC whenever needed. A moderate depth of anesthesia was maintained during each procedure by clinical judgment and occasional EEG monitoring.

TABLE 11–I
STEROID-TREATED PATIENTS

Case	*Age*	*Sex*	*Dosis/day (mg)*	*Duration (Days)*	*Total Dosis*† *(g)*	*Operation*	*Time Between* Last Dose and Operation*
1	24	F	119	7	1.08	Ventriculoatriostomy	0**
2	57	M	59	23	1.35	Choledochostomy	17
3	22	M	56	28	1.59	Splenectomy	2
4	59	F	—	—	—	Hysterectomy	30
5	48	M	45	50	2.26	Gastrectomy	180
6	43	M	60	20	1.20	Choledochostomy	0
7	57	M	41	297	12.27	Rectal amputation	7
8	36	F	313	234	732.46	Hysterectomy	0
9	17	M	58	86	5.00	Lymphadenectomy	12
10	45	M	84	70	5.88	Extirpation of tumor of parotid gl.	21
11	25	M	59	113	6.88	Pancolectomy	0
12	22	F	58	83	0.85	Oophorectomy	8
13	60	M	57	74	4.19	Choledochostomy	15
14	39	M	78	17	1.33	Ventriculoatriostomy	2
15	26	M	98	126	12.41	Exploratory abdominal incision	0
16	70	M	80	24	2.64	Choledochostomy	9
17	17	F	106	279	29.52	Splenectomy	0
18	68	F	56	71	3.91	Extirpation of mediastinal tumor	21
19	24	M	30	1855	55.65	Bone grafting	0

* Days after the last steroid administration prior to one day before operation.
** Continuous steroid administration prior to one day before operation.
† Equivalent dosis to cortisol.

The determination of free cortisol concentration was made for 1 ml of plasma by Rudd's fluorometric procedures with Farrand spectrofluorometer, which modified Silber-Porter method.[12] The blood sampling time was as follows: (1) immediately before induction of anesthesia, (2) every thirty to sixty minutes after halothane anesthesia, and (3) in the recovery room when the patient was considered to be adequately recovered from anesthesia.

Table 11-II and 11-III show the effect of halothane anesthesia and operation on the nineteen steroid-treated patients, compared with the twenty control patients. The response in the two groups was markedly different, showing significant lower levels of plasma cortisol in the

TABLE 11-II
PLASMA-FREE CORTISOL LEVELS (μg/100 ml) IN STEROID-TREATED PATIENTS DURING HALOTHANE ANESTHESIA AND SURGERY

Case No.	*Preinduction*	*30 min After Start of Anesthesia*	*Time After Start of Operation*			*Recovery Room*
			30 min	*1 hr*	*2 hr*	
1	7.3	11.1	15.7	14.8	22.8	28.6
2	4.2	0.0	—	20.0	17.9	24.0
3	4.3	13.5	—	—	13.5	—
4	15.0	15.5	20.0	18.5	23.5	25.0
5	3.3	6.1	20.0	21.1	33.3	27.8
6	16.2	13.7	9.3	6.2	10.6	9.3
7	5.8	8.1	11.8	12.6	15.5	20.7
8	1.0	2.5	—	7.0	—	8.5
9	18.3	—	17.9	12.2	22.0	39.5
10	17.5	15.0	11.3	11.3	11.3	7.5
11	18.0	14.0	9.0	9.0	11.0	14.8
12	4.5	7.0	—	11.5	26.0	28.5
13	14.2	14.3	8.3	8.3	8.0	6.5
14	12.5	11.3	—	3.8	5.0	27.5
15	17.5	20.0	21.5	22.5	—	—
16	15.0	17.5	—	27.5	—	—
17	5.8	14.5	7.2	10.0	—	18.5
18	8.0	10.0	12.0	20.0	22.0	—
19	13.0	5.0	22.5	7.5	20.0	22.0
Mean	10.6	11.1	14.3	13.5	17.5	20.6
SE	1.3	1.2	1.5	1.5	1.9	2.4
t		0.249	1.814	1.430	2.958	3.723
p		0.959	0.080	0.825	0.006*	0.001*

* Statistically significant compared with preinduction mean value.

TABLE 11-III
DIFFERENCES IN CORTISOL RESPONSES TO HALOTHANE ANESTHESIA AND SURGERY BETWEEN STEROID-TREATED PATIENTS AND CONTROL PATIENTS
(μg/100 ml)

	Time in Minutes					
Patients	*Preind.*	*Anes. 30*	*Op. 30*	*60*	*120*	*R.R.**
Control (20)	12.6 ± 0.9	17.5 ± 1.1	23.6 ± 2.5	27.0 ± 1.5	29.1 ± 2.1	30.6 ± 1.4
Steroid-treated (19)	10.6 ± 1.3	11.1 ± 1.2	14.3 ± 1.5	13.5 ± 1.5	17.5 ± 1.9	20.6 ± 2.4
t	1.209	3.856	3.263	6.198	3.915	3.681
p	NS	<0.001	<0.01	<0.001	<0.001	<0.001

± = Standard error of the mean.
* = Adequately recovered from anesthesia in the recovery room.
( ) = No. of patients.

steroid-treated patients than in the control patients during anesthesia, surgery, and postoperative period.[13]

Furthermore, in contrast with the control patients, no significant increase of plasma cortisol was noted in the steroid treated patients after thirty minutes of anesthesia before operating, nor one hour later after the start of the operation, in comparison with the preinduction level.

As I mentioned previously, it is quite interesting to note that despite the increasing number of steroid-treated patients recently, most of the so-called "adrenal failures" were reported between 1952 and 1958, although it has been known for many years that administration of corticosteroids can cause atrophy of the adrenal glands. Adrenocortical insufficiency during or after surgery has rarely been documented. Many cases of hypotensive patients who have had satisfactory rsponses to the administration of corticoids along with other treatments have been interpreted as establishing the diagnosis of adrenocortical failure with no proof other than the response to therapy. Many of these cases resembling superficial adrenal insufficiency recover after the correction of blood loss or injection of vasopressor.[14,15] By these reasons, it is of paramount importance to measure adrenocortical function during hypotensive episode in patients who have a history of receiving steroid therapy in the past or present.

In about half of the steroid treated patients in our series, the low plasma-free cortisol levels were noted at preinduction period in the morning on the day of surgery. The similar low levels remained to continue during halothane anesthesia alone for thirty to forty-five minutes.

However, we found six hypotensive patients whose systolic blood pressure decreased below 80 mm Hg. And in two of these six patients, the plasma cortisol levels were found not to be low. In these six hypotensive patients, only one needed administration of corticosteroid along with other treatments during operation to restore blood pressure.

Now I should like to present the interesting case of a fifty-seven-year-old patient who underwent rectal resection for rectal cancer. He had been under corticosteroid treatment, consisting of dexamethasone (1 mg/day), for one year. He took the estimated total equivalent dose of 12.2 gm of cortisol until a week prior to operation. The arterial blood pressure decreased from 150 to 80 mm Hg twenty minutes after surgery began, when the plasma cortisol level was very low, 11.8μg. Blood loss was approximately 50 ml at this time and vasopressor (effortil, 2 mg) was administered intravenously. Within twenty minutes arterial blood pressure elevated to 140 mm Hg.

The next case is a forty-three-year-old man who underwent choledochostomy for pancreatic cancer. He had received dexamethasone (2 mg per day for twenty days), the estimated total equivalent dose of 1,200 mg of cortisol until the day before operation. His blood pressure continued to decrease from 78 to 54 mm Hg, while plasma cortisol levels reduced to 10μg/100 ml after forty-five minutes of halothane anesthesia alone. Vasopressor (effortil, 2 mg) failed to restore the blood pressure, dexamethasone (2 mg) was intravenously injected and the blood pressure was increased to 120/80 mm Hg. It again decreased to 90/70 mm Hg two hours after the operation started, reflecting low plasma cortisol levels between 6μg and 9μg/100 ml, 2 mg of dexamethasone intravenous injection was again effective to restore blood pressure.

Valid comparisons are needed with respect to the magnitudes of the rise in plasma cortisol concentration in the steroid-treated patients and their blood pressure during anesthesia and operation. So far as the timing of hypotension is concerned, it is not necessarily true that the lowest plasma cortisol level corresponded well with the arterial hypotension in the given time. Moreover, we found four other patients whose plasma cortisol levels were less than these patients (0μg–7μg/100 ml) and did not develop hypotension.

The observed incidence of hypotension may appear low, and it seems reasonable to conclude from our observation that the low plasma cortisol level played a part in the development of hypotension in the steroid-treated patients. Nonetheless, the data do suggest that plasma cortisol levels are not the prime determinant of the blood pressure attained during operation in steroid-treated patients.[17-19] Our data would suggest that the low plasma-free cortisol level in the peripheral venous blood during anesthesia or surgery is not necessary to represent the arterial hypotension, and vice versa in the steroid-treated patients.

It has been widely recommended that patients receiving steroids within a two-year period before the operation should have steroid cover to prevent adrenal failures.[8] Vandam and Moore[16] stated that any patient who has been on cortisone therapy for four days or longer, or within the past six months, may be considered to have subnormal adrenocortical function.

Since adrenal insufficiency following steroid therapy leading to collapse during the operation is rare, it hardly seems justifiable to submit all patients to steroid cover, thus increasing the risk of infection and gastrointestinal bleeding, in order to safeguard the few with adrenal insufficiency. Thus a preoperative test is needed.

Unfortunately, in our institution the anesthesiologist is informed about the schedule of operation on the day before operation, therefore it is difficult to use ACTH test before operation. Nonetheless, for the safety of patients we must get the cooperation of the surgeons. Thus I was not able to use ACTH test on surgical patients, but I used it on several medical patients and found that the steroid-treated patients' responses were subnormal.

Another new and more convenient method of ACTH test is the injection of synthetic corticotropin (Synacthen®) intramuscularly. A rise of less than 7μg was designated as an "impaired Synacthen test." It is quite evident that the adrenocortical response to Synacthen in our patients was subnormal.

The response to severe stress calls for greater corticotropin release than does the restoration or maintenance of normal adrenal activity. Clearly, the stress response cannot be tested by observing the rate of recovery after weaning from steroids or by corticotropin stimulation, because the stress response is determined by a mechanism independent of that controlling normal plasma cortisol.[14,20]

Sampson *et al.*[15] reported that those showing an unsatisfactory response collapsed during operation, unless given cover. Jasani *et al.*[17] recommended that patients with subnormal responses to Synacthen require corticosteroid cover: one out of nine of such patients developed hypotension. A very important point worthy of further study is whether the patients with a normal response to Synacthen do not require corticosteroid cover and whether the patients with subnormal reactions should receive routine steroid cover.

Recommendations regarding corticosteroid cover during operations on corticosteroid-treated patients are still largely empirical. From the practical point of view, it is not always possible to apply ACTH test routinely to steroid-treated patients.

The following patients should have steroid cover:

1. The patient who is now under steroid treatment.
2. The patient who received steroid treatment for more than one month within the six months prior to operation.
3. The patient who had total steriod of more than 1000 mg of equivalent of cortisol within the six months prior to operation.
4. The patient who is to undergo adrenalectomy or hypophysectomy or who has a history of these operations.
5. Addisonian.

Our scheme of cortisone dosage during the operative period is to give two doses of 100 mg hydrocortisone I.M., one at the time of premedication, and the other during the operation by I.V. drip. Following major surgery, this regime should be continued for three days; for minor operations, twenty-four hours is adequate.

Now I should like to mention the surgical and medical complications by corticosteroid treatments. The most important for anesthesiologists is the bleeding tendency. In our series, because of the heavy bleeding (about 300 ml), the collectomy was stopped before intraperitoneal maneuver started.

Main surgical and medical complications are shown in Tables 11-IV and 11-V. However, we should not forget the benefit from corticosteroids to surgical patients. The indications for them are listed in Table 11-VI.

The effect of corticosteroids to unresponsive shock is dramatic once in a while. The mechanism of it is suggested as follows: improvement of

TABLE 11-IV
SURGICAL UNTOWARD EFFECTS OF PROLONGED STEROID THERAPY

1. Wound infection	5. Severe electrolyte imbalance
2. Wound dehiscence	6. Intra-abdominal abscess
3. Hemorrhage (G.I.)	7. Psychosis
4. Septicemia	

microcirculation, increase in production of the energy component ATP, stimulation in conversion of lactic acid to glycogen, protection of lysosomes, and increase in cardiac output. I have observed cases of hypovolemic shock which were improved by blood transfusion and by large dosage of paramethasone.

As a conclusion, I should like to reemphasize the fact that surgical hypotension due to the prolonged steroid therapy without steroid cover is rare, and the routine steroid cover for these patients is not recommended.

TABLE 11-V
MEDICAL UNTOWARD EFFECTS OF PROLONGED STEROID THERAPY

1. G.I. ulceration and hemorrhage	6. Osteoporosis
2. Susceptibility to infection	7. Muscle weakness
3. Diabetes	8. Acute psychosis
4. Electrolyte imbalance	9. Pancreatitis
5. Hypertension	10. Retardation of growth in children (?)

Finally, I should like to mention briefly interesting findings which were made at our laboratory. Levels of ACTH in the peripheral plasma of ten patients were determined during ether anesthesia alone and during and after laparotomy by a modification of Lipscomb and Nelson's bioassay method.

During induction of anesthesia and during surgery, large amounts of ACTH were released intermittently, with two or three steep peaks in the plasma levels. In contrast, the level of free cortisol in the plasma

TABLE 11-VI
INDICATIONS FOR USES OF ADRENAL STEROIDS IN ANESTHESIA AND SURGERY

I Primary Indications
1. Adrenalectomy
2. Hypophysectomy
3. Pituitary adrenal insufficiency
4. Cerebral edema
II Secondary Indications
1. Unresponsive shock
2. Some hemorrhagic states
3. Asthmatic attacks
4. Allergic emergencies
5. Glottic edema
6. Incompatible transfusion reactions
7. Aspiration pneumonia
8. Gram-negative septicemia
9. After organ transplantation

rose gradually. Our data indicate that the increased levels of free cortisol in the plasma were the results of increased plasma ACTH activity, but were not due to the impairment of cortisol metabolism. Furthermore, it would deny the feedback mechanism during stress.

We have studied adrenocortical activity in 329 patients. And all anesthetic agents are not necessary to stimulate the adrenocortical function.

For many years, premedication has been considered to have no special effects on adrenocortical activity. But, we recently observed that the premedicating drugs are capable of reducing the rise of plasma cortisol levels due to preoperative emotional stress. These are well corresponded with subjective evaluation.

Hypothermia is not necessary to depress stress response completely as judged from peripheral plasma cortisol levels during ether anesthesia, but it is necessary during methoxyflurane anesthesia.

It has been believed that spinal and epidural anesthesia can block adrenocortical response to stress. These techniques failed to block the elevation of plasma cortisol levels caused by the operation even during minor operations, although the magnitude in the increase of plasma cortisol level is low.

REFERENCES

1. Fraser, C.G., and Preuss, F.S.: Adrenal atrophy and irreversible shock associated with cortisone. *JAMA, 149:*1542–1543, 1962.
2. Salassa, R.M.; Benett, W.A.; Keating, F.R., and Sprague, R.G.: Postoperative adrenal cortical insufficiency occurrence in patients previously treated with cortisone. *JAMA, 152:*1509–1515, 1953.
3. Lewiss, L.; Robinson, R.F.; Yee, J.; Hacker, L.A., and Eisen, G.: Fetal adrenal cortical insufficiency precipitated by surgery during prolonged continuous cortisone treatment. *Ann Internal Med, 39:*116–126, 1953.
4. Slaney, G., and Brooke, B.N.: Postoperative collapse due to adrenal insufficiency following cortisone therapy. *Lancet, 1:*1167–1170, 1957.
5. Harnagel, E.E., and Kramer, W.G.: Severe adrenocortical insufficiency following joint manipulation. *JAMA, 158:*1518–1519, 1957.
6. Hayes, M.A., and Kushlan, S.O.: Influence of hormone therapy for ulcerative colitis upon course of surgical treatment. *Gastroenterology, 30:*75–84, 1956.
7. Allanby, K.D.: Deaths associated with steroid-hormone therapy, an analysis of 18 cases. *Lancet, I:*1104–1110, 1957.
8. Dundee, J.W.: Anaesthesia and surgery in adrenocortical insufficiency. *Br J Anaesth, 29:*166–180, 1957.
9. Nicholas, J.A.; Wilson, P.D., and Umberger, C.J.: Induced hypoadrenalism in patients requiring orthopedic surgery. *JAMA, 164:*261–265, 1957.
10. Winstone, N.E., and Brooke, B.N.: Effects of steroid treatment on patients undergoing operation. *Lancet, I:*973–975, 1961.
11. Sampson, P.A.; Brooke, B.N., and Winstone, N.E.: Biochemical confirmation of collapse due to adrenal failure. *Lancet, I:*1377, 1961.

12. Rudd, B.T.; Cowper, J.M., and Crawford, N.: The determination of plasma free hydrocortisone and corticosterone by a combined fluorimetric and modified Porter-Silber procedure. *Clin Chim Acta, 6:*686–695, 1961.
13. Oyama, T.; Shibata, S.; Matsumoto, F.; Takiguchi, M., and Kudo, T.: Effects of halothane anaesthesia and surgery on adrenocortical function in man. *Can Anaesth Soc J, 15:*258–266, 1968.
14. Cope, C.L.: The adrenal cortex in internal medicine (1). *Br Med J, 2:*847–853, 1966.
15. Sampson, P.A.; Winstone, N.E., and Brooke, B.N.: Adrenal function in surgical patients after steroid therapy. *Lancet, II:*322–325, 1962.
16. Vandam, L.D., and Moore, F.D.: Adrenocortical mechanism related to anesthesia. *Anesthesiology, 21:*531–552, 1960.
17. Jasani, M.K.; Freeman, P.A.; Boyle, J.A.; Reid, A.M.; Direr, M.J., and Buchanan, W.W.: Studies of the rise in plasma 11–hydroxycorticosteroids (11–OHCS) in corticosteroid-treated patients with rheumatoid arthritis during surgery: correlations with the functional integrity of the hypothalamo-pituitary-adrenal axis. *Quart J Med, 37:*407–421, 1968.
18. Plumpton, F.S.; Besser, G.M., and Cole, P.V.: *Corticosteroid treatment and surgery Anaesthesia, 24:*3–18, 1969.
19. Mattingly, D., and Tyler, C.: Plasma 11–hydroxy corticoid levels in surgical stress. *Proc R Soc Med, 58:*1010–1012, 1965.
20. Oyama, T.; Saito, T.; Isomatsu, T.; Samejima, N.; Uemura, T., and Arimura, A.: Plasma levels of ACTH and cortisol in man during diethyl ether anaesthesia and surgery. *Anesthesiology, 29:*559–564, 1968.

Chapter 12

Fluid and Electrolyte Changes in Endocrine Disease and Their Significance in Anesthesia

M.T. Jenkins and A.H. Giesecke, Jr.

These changes, like so many of concern to the anesthesiologist, include the broad panorama of medicine. As influenced by anesthesia, fluid and electrolyte responses are not constant but present kaleidoscopic patterns and are not subject to precise interpretation. In considering fluid and electrolyte changes in endocrine disease, the anesthesiologist should first have a concept of the effects of anesthetic agents on renal function, and he should know the normal renal responses to a diminished circulating fluid volume.

A traditional view, considered now to be fallacious in view of clinical practices, is that renal function is altered specifically by general anesthesia, causing oliguria or anuria. Undeniably, total renal blood flow may vary according to the degree of hypotension produced. However, earlier theories that anesthetic agents per se stimulate production of antidiuretic hormone or of aldosterone do not seem tenable in view of clinical experience. Urine output continues during surgery if the extracellular fluid volume is maintained and bolstered by balanced salt solutions to compensate for that which is translocated incident to the patient's previous trauma or to the manipulations in the surgical procedure.

HOMEOSTASIS IN THE NORMAL UNANESTHETIZED PATIENT

The water and electrolyte composition of the body remains relatively constant despite extreme variations in external circumstances and environment. Physiological mechanisms exist to help in conservation of volume of fluid when in deficit and their excretion when in excess.

Normally, when the ECF volume is expanded as a result of high sodium intake or the action of sodium-retaining steroids, glomerular filtration increases, and there is a concomitant suppression of renin, angiotensin, and aldosterone production.

Depletion of ECF resulting from restricted sodium intake or translocations in the body (edema, peritonitis, bowel obstruction, fractures) leads to an increased production of renin, angiotensin, and aldosterone and results in a decreased glomerular filtration.

Antidiuretic hormone levels are increased if there is increased osmolality of body fluids which may occur when water loss exceeds water intake. As this results in a decreased circulating fluid volume, there is also a pronounced diminution in glomerular filtration. Both factors serve to conserve free water.

An excess water balance causes a decreased osmolality of circulating body fluids and consequently suppresses secretion of ADH. Glomerular filtration increases, and free water excretion is facilitated.

The concentration of potassium ions in the body normally is adjusted with considerable precision between cells and extracellular fluid. Abnormalities are life threatening. Aldosterone production is stimulated by high serum levels of potassium which accelerate excretion of potassium by the kidney and retention of sodium.

NORMAL RENAL FUNCTIONS IN THE REGULATION OF FLUIDS AND ELECTROLYTES

The antidiuretic hormone (ADH) of the posterior pituitary gland and aldosterone, the sodium-retaining hormone from the glomerulosa cells of the adrenal cortex, are the most important hormones in regulating water and electrolyte volume and metabolism under normal physiologic circumstances. They exert their principal effects on renal function to produce an organized sequence of events.

For the initial step in the formation of urine, the glomerulus functions somewhat as a spill valve, normally filtering 150 to 180 liters of water containing approximately 20,000 mEq sodium and 14,000 mEq chloride. Indicative of the rather precise balance between filtration and reabsorption is the fact that approximately 99 per cent of the filtered water, sodium, and chloride is reabsorbed in the renal tubules. The actual filtering at the glomerulus is probably nonhormonal. However, increases in the GFR, reflecting the increased renal blood flow, may result from increases in the levels of thyroid growth hormone and glucocorticoids. GFR will be increased, also, if the sodium-retaining hormones and ADH expand the extracellular fluid compartment.

The glomerular ultrafiltrate contains all of the crystalloid components in about the same concentrations as they exist in the water phase of plasma from which the filtrate is formed. As the glomerular filtrate traverses the proximal tubules, 80 to 87 per cent of the filtered sodium and chloride are actively reabsorbed. Bicarbonate ions may be reabsorbed indirectly by the passive migration of hydrogen ions from the tubular cell into the lumen of the proximal tubule. The osmotic force created by the active reabsorption of the ions and the other solutes accounts for the passive transfer of the similar volume of water. Whether hormones play a part in the transfer at the proximal convoluted tubule is not known.

Isotonic urine is delivered from the proximal convoluted tubule to the descending limb of the loop of Henle, and it becomes increasingly hypertonic until it is delivered into the ascending tubule. Sodium and chloride are actively transported out of the water-permeable ascending loop of Henle into the interstitial tissue of the renal medulla, establishing a gradient of about 200 mOsm/kg water between the interstitium and the lumenal fluid of the ascending limb. Water comes also from the descending limb of the loop of Henle into the interstitium, drawn by the increased osmolality caused by the electrolytes in the interstitium. Sodium reenters the descending limb perhaps by process of passive diffusion.

The handling of the tubular fluid is believed to be the same in both water diuresis and antidiuresis up to the beginning of the distal renal tubule. Active sodium reabsorption continues in the distal tubule. In the absence of ADH, the distal convoluted tubule is relatively impermeable to water. The distal convoluted tubule reabsorbs sodium and chloride in a process which involves the secretion of hydrogen and potassium ions into the tubular lumen. Ammonia, which is synthesized by the tubular epithelium, diffuses into the lumen. In the presence of hydrogen it is converted into the ammonium ion. These processes are all stimulated by aldosterone-like steroids and are inhibited by aldosterone antagonists. With the maximum ADH activity a comparatively low volume of isotonic fluid is delivered into the collecting duct.

Water and other small molecules, such as urea, are passively reabsorbed from the collecting duct. In the absence of ADH it is relatively impermeable to these substances. ADH increases the permeability of the collecting duct to water. Under the influence of ADH, therefore, a small volume of hypertonic urine is finally excreted, usually about two liters a day from the original 180 liters of glomerular filtrate.

THE SPECIFIC EFFECTS OF HORMONES ON FLUID AND ELECTROLYTE METABOLISM

Antidiuretic Hormone

Antidiuretic hormone increases renal tubular permeability to water, permitting water reabsorption by distal convoluted tubules and collecting ducts to expand the total body water compartment. This causes a dilution of intracellular and extracellular electrolytes, depresses aldosterone secretion, and leads to increased sodium and decreased potassium excretion.

ADH is secreted in response to dehydration and its ultimate effect is to overcome dehydration. Conversely, a decrease in the osmolality of ECF below approximately 280 mOsm/kg water normally leads to cessation of ADH secretion. Clinical disorders of ADH physiology include diabetes insipidus and hypovolemia. One form of inappropriate ADH excess has been found in the production of ectopic ADH by certain tumors of the lung. Water intoxication can develop in these cases, since hypertonic urine is excreted even though serum osmolality is distinctly subnormal. In nephrogenic diabetes insipidus, a familial disorder, there is a limited responsiveness to ADH. Patients having hypokalemia or hypercalcemia exhibit a kidney dysfunction also in that the kidney loses its ability to generate a normal degree of hypertonicity of the medullary interstitial fluid.

Aldosterone

Aldosterone causes an increased cation exchange in the distal convoluted tubules of the kidney, specifically increasing urinary excretion of potassium and titratable acid and ammonia and decreasing excretion of sodium. This expands the extracellular fluid compartment and results in hypertension, alkalosis, edema, polyuria, and polydipsia.

In the distal convoluted tubule, aldosterone acts to increase the reabsorption of sodium and chloride from the tubular lumen and to increase the secretion of potassium and hydrogen ion into the lumen. The mechanism regulating the secretion of aldosterone is separate from that controlling the secretion of the other major adrenal steroids. Aldosterone is appropriately responsive to changes in body hydration. The question of what regulates the secretion of aldosterone is resolved into a question of what controls the secretion of renin. Conditions that reduce extracellular fluid volume, that reduce total blood volume, and that result in the formation of renin by the juxtaglomerular apparatus thereby increase the formation of angiotensin. ACTH has a transient stimulatory effect on the secretion of aldosterone. Hypovolemia stimulates ten times more aldosterone than does ACTH. Aldosterone secretion is also influenced by the concentration of potassium in body fluids. Diminished potassium decreases the secretion of aldosterone, whereas an excess of potassium stimulates the production of aldosterone as a potassium-wasting hormone.

Cortisol

Cortisol increases renal tubular cation exchange and increases excretion of water, potassium, ammonium, and titratable acids while retaining sodium. Adrenal cortical hormones increase filtration at the glomerulus.

Progesterone

Progesterone (and spironolactones, which are not true hormones) exert competitive inhibition of aldosterone at the renal tubular level to cause an increased excretion of sodium and chloride and water and a decreased excretion of potassium and ammonium and titratable acid. These compounds have little effect on fluid or electrolyte metabolism when given in the absence of aldosterone-like steroids.

Thyroid Hormones

Thyroid hormones increase cardiac output and renal blood flow and GFR, causing an increased excretion of water and electrolytes.

Growth Hormones

Growth hormones promote tissue anabolism with a retention of intracellular and extracellular water and electrolytes in quantities appropriate for growth and development.

Androgens

Androgens function like growth hormones plus aldosterone to cause retention of intracellular and extracellular water and electrolytes.

Parathyroid Hormone

Parathyroid hormone promotes reabsorption of calcium salts from bone, resulting in hypercalcemia and hypercalciuria. There is no significant regulation of other electrolytes unless nephrocalcinosis results from the hypercalcemia. If this occurs, it may indirectly cause a renal tubular disorder leading to hyposthenuria, dehydration, polyuria, and polydipsia.

Insulin

Insulin causes no direct effect upon water and electrolyte metabolism. In insulin deficiency, however, hyperglycemia can cause an osmotic diuresis promoting loss of water and electrolytes in the urine. If acidosis develops because of ketone body formation, there will be an increased excretion of hydrogen ion and ammonium ion plus an increased excretion of fixed cations such as sodium and potassium.

CHANGES IN VARIOUS ENDOCRINE DISORDERS

Obesity

Whether the patient's physique is a matter of habit or hormones, it still is logical to discuss the anesthetic problems in relation to fluids and electrolytes. Since the obese patient has a smaller total body water content as compared with the nonfat individual of similar weight, he has a reduced ability to compensate for fluid losses from the body or fluid translocations within the body. For the fat patient, his total body water may be only 35 to 50 per cent of body weight in contrast to the 60 per cent body weight for the nonfat adult male (73 per cent of lean body weight in either case). The amount of water in different tissues of the body varies greatly: 80 to 85 per cent in the gray matter of the brain; 70 to 78 per cent in muscle; 14 to 44 per cent in bones; 6 to 10 per cent in fat; 3 per cent in tooth enamel.

Weisberg has devised a practical and simple equation for determining total body fat: % fat = 90–2 (height in inches minus girth in inches). By modification, the anesthesiologist may use this in estimating the total ECF which, with the blood volume, constitutes the circulating volume.

Extravascular ECF = .20 × total body water
Total body water (as % of body weight) =
    0.732 × lean body mass
Lean body mass = 100 − % fat
% fat = 90 − 2 (height in inches, girth in inches)
Therefore: Total body water =
    0.732 (100 − [90 − 2 (H − G)]) = 7.32 + 1.46 (H − G)

Example: Male, 70 kg, 5′9″, 33″ waistline
Body water = 7.32 + 1.46 (69–33)
= 7.32 + 1.46 (36)
= 7.32 + 52.56
= 59.88% body weight

Total body water = 42 liters
ECF = 8.4 liters

Example: Male, 100 kg, 5′9″, 44″ waistline
Body water = 7.32 + 1.46 (69–44)
= 7.32 + 1.46 (25)
= 7.32 + 36.50
= 43.82% body weight

Total body water = 44 liters
ECF = 8.8 liters

These calculations are of value primarily to emphasize that the obese patient has a relatively smaller ECF volume and will consequently have less ability to withstand the stress of a given volume external fluid loss or internal translocation as compared with the nonfat patient.

Diabetes Mellitus

Statistically, diabetes mellitus is the most commonly encountered endocrine disease, occurring in 1 to 2 per cent of the population. Therefore, it is the most frequent endocrine problem confronting the anesthesiologist, for the diabetic patients are subject to all the surgical diseases of their particular age groups and are further subject to many surgical conditions which are secondary to their endocrine imbalance.

Regardless of frequency of appearance of diabetic patients on the operative schedule, there does not exist a unanimity of opinion among anesthesiologists, surgeons, and internists about their preoperative management. Generally we seem to "get away with" whatever regimen of care is followed, probably because 75 per cent of the patients have the adult-onset type which is ketoacidosis-resistant. Of greater significance and with a smaller margin of safety are the juvenile diabetic patients. Management of all diabetic patients during anesthesia can best be accomplished if no preoperative insulin is given. The anesthesiologist can then utilize simple tests to determine blood sugar and acetone at intervals during the operation.

In considering the serum electrolyte values reported for the diabetic patient in hyperglycemia and acidosis, remember that a correction factor must be applied to a low serum sodium reading to know the effective osmolality. By its osmotic effect a significant hyperglycemia will draw fluid into the extracellular compartment and thus decrease the concentration of electrolytes therein. Also, by its molecular size and presence, glucose will "displace" a number of sodium ions so that in a measured volume of serum there will be a hyponatremia though a normal effective osmolality. The following factors are important in calculating the true situation, as described by Welt:

A. Measured glucose minus 100 mg/100 ml (normal value) times 10 (mg excess per liter) divided by 180 (mol wt glucose) equals excess glucose mOsm/liter displacing NaCl.

B. Excess glucose mOsm/liter divided by 2 equals excess glucose mOsm/liter displacing sodium ions. (Divide by 2, since 1 mM of sodium *salts* provides 2 mOsm.)

C. Add this number excess glucose to measured sodium concentration for correct mEq/liter value of sodium.

*Example:**

Measured blood sugar = 1,000 mg/100 ml
Measured serum sodium = 115 mEq/liter

* These steps may be simplified by dividing the excess glucose by the factor of 36 to determine the figure to add to the actually measured serum sodium.

$1,000 - 100 = 900 \times 10 = 9,000$ mOsm glucose/liter

$\frac{9,000}{180} = 50 \qquad \frac{50}{2} = 25$

$115 + 25 = 140$ mEq/liter sodium

The tonicity due to sodium will be normal when hyperglycemia is corrected, and consequently other special efforts to replace sodium would cause a hypernatremia.

Generally we feel that the patient with adult diabetes should have a blood sugar of 150 to 200 mg/100 ml with a negative serum acetone before the induction of anesthesia. The patient with juvenile or ketoacidosis-prone diabetes can best be protected if his preanesthetic blood sugar is above 200 mg/100 ml with a negative serum acetone.

Diabetes Insipidus

The neurohormonal unit responsible for ADH (vasopressin) production and secretion includes the supraoptic nuclei, the supraopticohypophysial tract, and the posterior pituitary or neurohypophysis itself. There are at least two routes of stimulation of ADH by the neurohypophysis. One is through a neuronal pathway leading to activation of hypothalamic nuclei by noxious agents, by psychic stress, and by drugs such as acetylcholine and nicotine. The other mechanism is by way of stimulation of osmoreceptors somewhere within the distribution of the internal carotid arteries by alterations in the osmotic pressure of the plasma.

Diabetes insipidus is usually caused by lesions of the hypothalamic tract or by lesions which interrupt the supraopticohypophysial tract. The anesthesiologist may be confronted with diabetes insipidus in the patient who has been the victim of trauma, especially that involving basilar skull fractures. It may also follow meningitis or encephalitis or be caused by a metastatic lesion or by granulomatous disease. Damage to the posterior pituitary gland sufficient to destroy at least 95 per cent of the gland and extend up the neural stalk into the median eminence of the hypothalamus will also cause diabetes insipidus. Hypophysectomy itself (unless the hypothalamus is inadvertently damaged during the operative procedure) results in a mild diabetes insipidus usually for only two or three days' duration, since ADH continues to be produced by neurons of the supraoptic nuclei and the storage function of the posterior pituitary gland may be taken over by the hypothalamus. Nephrogenic diabetes insipidus is a rare hereditary disorder and there are other renal lesions which may cause a picture similar to nephrogenic diabetes insipidus. This may occur in the recovery phase of acute tubular necrosis after the sudden relief of ureteral obstruction, in hypercalcemia, and in hypokalemic nephropathy. In all of these circumstances, damaged renal tubular cells show an inadequate response to antidiuretic hormone so that water is not reabsorbed in the distal convoluted tubule and in the collecting tubules.

The basic defect in diabetes insipidus is an inadequate supply of ADH. It is thought that the hormone vasopressin produced by the neurones of supraoptic nuclei migrate along the nerve fibers to be stored in and possibly released by the posterior lobe of the pituitary gland. The supraoptic nuclei respond to the osmotic pressure of the blood and release the amount of ADH necessary to maintain normal water balance. ADH carried to the kidneys exerts an effect on the distal convoluted tubules and the collecting ducts. Thus, when the hypothalamus is bathed in hyperosmotic blood, as in dehydration, the amount of ADH rises and acts on the distal tubules of the nephron to decrease the urinary volume. Conversely, in overhydration, the concentration of ADH is reduced, permitting an increase in urinary volume.

In diabetes insipidus, with the loss of ADH and its influence on the kidney, the reabsorption of fluid fails to take place and results in polyuria. Adults have been known to excrete up to forty-three liters of urine per day, although this is a greatly exaggerated volume. Children may excrete five to ten liters per day. The only resource the untreated patient has for compensation is increased water intake. Therefore, the primary symptoms are polydipsia and polyuria. Total salt excretion is not increased, and only rarely are there nutritional elements lost in excess of the normal excretion in the urine. For the patient not on replacement therapy, the danger is loss of access to fluids, since with a diminished intake and failure of the kidneys to concentrate urine, dehydration and hypernatremia will soon result.

We have recently had a four and one-half-year-old boy who was admitted for surgical excision of an eosinophilic granuloma of the gum. He had received general anesthesia two years previously for excision of an eosinophilic granuloma of the left frontal bone, after which he had received x-ray therapy to localized areas of the skull. During the interim between these two admissions, he had developed polydipsia and polyuria and a Pitressin® test indicated a diagnosis of diabetes insipidus. His thirst and polyuria were controlled after this by nasal insufflation of posterior pituitary powder every eight hours.

In considering his preparation for surgery, the question was debated whether to continue his Pitressin, administering the last dose immediately prior to induction, or whether to discontinue it eight hours preanesthetic and cope with urine output by fluid administration. Either alternative in his preanesthetic preparation posed some problems for the anesthesiologist. First, studies on ADH in experimental animals have revealed severe myocardial depression due to the compound. A transient tachycardia followed by prolonged and profound bradycardia may occur, simulating the syndrome of myocardial hypoxia due to coronary insufficiency. Conceivably, in his chronic usage of the nasal insufflation of the posterior pituitary powder, this patient could develop a resistance to the effects which are shown to occur in the acute experimental animal. However, the anesthesiologist may be concerned over whether the coronary effects will supervene under general anesthesia. As an alternative to be considered by the anesthesiologist, the vasopressin could be discontinued eight hours prior to beginning of anesthesia so that the patient would be in an untreated state of diabetes insipidus during anesthesia. Lacking experience in anesthetizing patients with diabetes insipidus but realizing the possible bad effects under anesthesia of myocardial ischemia subsequent to vasopressin, we elected to discontinue the antidiuretic agent eight hours before beginning anesthesia and allowed the patient water ad lib by mouth until one hour prior to induction of anesthesia.

The pros and cons of this decision exceed the limit of the present discussion, since there is a question whether the anesthesiologist could easier handle the problems of myocardial ischemia and coronary insufficiency secondary to vasopressin-Pitressin injection or could maintain proper fluid balance in the case of urine output of two or more liters per hour.

Ploss and Warner report a difficult problem in the anesthetic and postanesthetic management of a patient with juvenile ketoacidosis-prone diabetes compounded by the acute development of diabetes insipidus. Their patient was a twenty-six-year-old man with a history of sixteen years of insulin-treated diabetes mellitus who had developed a diabetic retinopathy. In the operative procedure the surgeons encountered great difficulty in exposing the pituitary stalk, and retractor pressure was applied to the hypothalamus posterior to the stalk bilaterally. On the first postoperative day, this patient excreted forty-nine and one-half liters of urine. During the following days, this patient had urinary outputs of thirty-eight liters, thirteen liters, forty-six liters, ten liters, and the seventh day he was anuric. His serum glucose ranged from 1200 mg/100 ml on the second day down to 188 mg/100 ml. Hypokalemia was a repeated problem, and there was one brief period of hyponatremia.

For such a case as this, the usual anesthetic intraoperative problems involve primarily positioning, airway maintenance, ventilation, monitoring of physiological variables, problems concerning brain decompression, and blood replacement. Fluid administration posed more difficult problems. For the postoperative period, also, there are many endocrine sequelae. Adrenal insufficiency may occur rather suddenly and acute signs develop within thirty-six to seventy-two hours after operation. Warner and Ploss noted that the normal adrenal secretes 20 to 25 mg hydrocortisone in twenty-four hours, and since anesthesia and surgery cause a two to threefold increase in steroid output, they recommend providing 75 to 100 mg of hydrocortisone for twenty-four hours beginning on the day of surgery. Upon loss of trophic hormone after a pituitary stalk section, insulin tolerance in the diabetic patient is usually promptly reduced and the hypophysectomized patient rarely has postoperative ketoacidosis. Nonketotic hyperglycemic hyperosmolar coma, however, may follow pituitary stalk section. During a long operative procedure, it becomes incumbent upon the anesthesiologist to follow serum glucose and acetone to monitor the status of the diabetic patient. Concerning the development of diabetes insipidus following stalk section, it is usually noted that a section close to or below the level of the diaphragma sellae is followed by a very minor period of diabetes insipidus, during which there may be nearly normal urine osmolality and little polyuria. However, if the stalk is sectioned higher or if there is moderate or severe damage to the osmoreceptors in the hypothalamus, a permanent diabetes insipidus may result. Profound polyuria developing after stalk section, as reported, may in part be due to an untreated hyperglycemia as well as to the loss of vasopressin.

Despite difficult problems with this one case, Ploss and Warner state that the conscious patient with diabetes insipidus can usually be kept in adequate fluid balance by simply replacing urinary loss with sufficient fluid volume. They note that the average adult patient with severe diabetes insipidus does not usually excrete more than six to ten liters of urine daily, and this amount may be lessened by reducing the protein and salt intake. For their reported patient, the large urine volumes may have resulted from excess administration of fluids plus the persistent hyperglycemia which was not completely corrected. The continuing hyperglycemia increased the osmotic load presented to the kidneys, augmenting the polyuria. Also, the hyperosmolality of the serum promoted loss of intracellular fluid and further impaired this patient's CNS function.

Perhaps the problems of diabetes insipidus developing subsequent to pituitary stalk section could be minimized by limiting the patient's salt intake to 5 gm per day, keeping the total fluid administration or intake below six liters and controlling the blood sugar level to a range of 150 to 200 mg/100 ml. If the patient's urine output shows signs of exceeding the fluid administration, aqueous preparations of ADH may give hour-to-hour control. Since aqueous ADH is expected to act for only three to five hours, it may be advisable to switch to a longer acting ADH, vasopressin tannate, after the first week. Vasopressin tannate is suspended in vegetable oil which should be warmed and vigorously shaken before administration every thirty-six to seventy-two hours. For persistent polyuria and low urine osmolality beyond three months postoperatively, indicating permanent diabetes insipidus, the patient may add nasal insufflation of posterior pituitary powder as chronic treatment in addition to replacement with cortisone and thyroid and sex hormones.

Pheochromocytoma

As reported by Crout and Brown, surgical mortality in pheochromocytoma is now relatively low, although it was reported to be as high as 25 to 45 per cent in the pre-1950 era. For the decade 1951–60 there is a documented overall figure of about 13 per cent, and papers since 1960 suggest that the current operative mortality is much less, perhaps 1 to 5 per cent. The increasing success in surgical and anesthesia management may be due to better diagnostic methods, more effective drugs for control of hypertensive episodes, and an increased awareness of hypovolemia which exists in patients with pheochromocytomas.

Crout and Brown present evidence from experience with fourteen cases that the incidence of cardiovascular crises is relatively low in patients who have been treated prior to surgery with oral phenoxybenzamine (Dibenzyline®) who are then anesthetized with methoxyflurane (Penthrane®). These patients are reported to come to surgery symptom-free and with blood pressure under control and with an adequate plasma volume. Blood pressure spikes during surgery were found to be blunted but not abolished, and their patients had little tendency to develop cardiac arrhythmias. The development of postoperative hypertension was not a problem.

The general assessment of the patient with pheochromocytoma is that he has a blood volume which is usually below normal, reflecting the general vasoconstricted state of the vasculature as influenced by the increased catecholamine elaborations of the tumor. (This low blood volume is in marked contradistinction to another patient with hypertension, specifically the patient with hyperaldosteronism where blood volume is elevated.) Crout and Brown found that the administration of phenoxybenzamine for several days prior to surgery permits the physiological correction of the plasma volume, and in addition provides a simple means of detecting those patients who also have a diminished red cell mass. Patients who demonstrate a sharp fall in hematocrit as a result of phenoxybenzamine therapy may be properly assessed as having a very low blood volume and should be transfused preoperatively.

The doses of phenoxybenzamine used in the pretreatment of the patient with pheochromocytoma should be such as to produce only partial adrenergic blockade. This will leave intact typical signs of diagnostic usefulness to the surgeon, including a finding of a blood pressure increase during manipulation of the tumor, and a blood pressure fall when the tumor is removed. The incidence of postoperative hypotension appears to be, if anything, less than phenoxybenzamine-treated patients and when hypotension does occur it can be controlled usually by rapid volume expansion with a balanced salt solution.

Primary Hyperaldosteronism

The aldosterone-secreting tumor is a bizarre endocrine dysfunction which poses potential problems for the anesthesiologist. Derangements are reflected into many systems, notably into electrolyte changes, and hypertension is found in nearly all patients. With the increased secretion of aldosterone from the glomerulosa cells of the adrenal cortex, hypernatremia without edema results from the increased water and sodium reabsorption. Symptoms are produced by potassium depletion as there is excessive loss of potassium and hydrogen ions through urinary excretion. The symptoms of potassium depletion usually appear as fatigue and loss of stamina. As the hypokalemic alkalosis progresses, the patient may exhibit a positive Chvostek and a positive Trousseau sign. These signs in an untreated hypertensive patient form presumptive evidence of alkalosis and possibly of hyperaldosteronism. Renal tubular dysfunction occurs also and is manifested by polyuria and polydipsia and, in fact, there is poor concentrating ability after dehydration or unresponsiveness of the kidney to ADH administration. As potassium and hydrogen ions are excreted in the urine, there is an increased sodium and water reabsorption. Cardiac problems during anesthesia will be life-threatening unless the potassium losses are replaced prior to operative exploration. For an emergency operation on a patient with untreated hyperaldosteronism, the anesthesiologist needs guidelines for rapidity of potassium administration.

(The total body potassium content normally is about 50 mEq/kg and all but about 3 per cent of this is in cells. Symptoms, however, reflect changes in serum potassium usually and initial treatment must be based on serum levels which should be closely followed.)

In emergency replacement where a large volume of potassium chloride is deemed necessary to avoid cardiac complications under anesthesia, the patient should be under close supervision with electrocardiographic monitoring to avoid hyperpotassemia. As an upper limit, if the patient is closely monitored, it may be safe to begin administration of 100 mEq potassium chloride in 1,000 ml 5% D/W, with the addition of twenty units of regular insulin. (The insulin aids the cellular uptake of potassium ions.) This solution may be continued until serial determinations show a serum potassium above 3 mEq/liter, which may be judged acceptable before beginning anesthesia.

Cushing's Disease or Cushing's Syndrome

When the various functions of cortisol or cortisone—the most abundant glucocorticoids—are deranged such as with exaggerated high levels, the resultant state will be Cushing's disease. Among the effects of cortisol is the increased glomerular filtration causing water diuresis with an increased excretion of potassium and hydrogen and a retention of sodium. Perhaps secondary to the sodium retention and water reabsorption is a moderate hypertension. Retention of sodium at the renal tubular level does occur because of the cortisol, but it is far less in magnitude than the sodium retention exerted by aldosterone. Cortisol excess leads to potassium depletion, causing alkalosis. Cortisol has a direct antagonistic effect to ADH in the collecting tubule, and so its ultimate effect is diuresis.

Because of other biological action of cortisol, the patient with Cushing's syndrome produces the additional findings of muscular weakness, obesity with a pendulous abdomen showing red striae, and abnormal fat distribution characterized by the establishment of a moon face, a pendulous abdomen, and fat pads supraclavicular and over the seventh vertebra. The latter distribution produces the so-called "buffalo hump" or "dowager's hump." Characteristically the arms and legs look thin. Severe osteoporosis may develop because of decreased deposition of bone matrix and a continuous loss of calcium in the urine.

From the standpoint of anesthesia, usually neither the hypokalemia nor the alkalosis resultant from Cushing's syndrome is sufficiently severe to require correction before surgery. However, regardless of whether the surgical approach is to remove a tumor of anterior pituitary or whether it is removal of an adrenal because of adenoma or carcinoma of the adrenal cortex, the prime concern will be in replacement of cortisone during the operative procedure. Following hypophysectomy, the adrenal cortex shows complete degeneration of the zona reticularis and fasciculata, the areas in which corticosteroids, estrogens, and androgens are produced. Consequently the patient in the postoperative phase will exhibit a chronic need for ACTH.

After unilateral adrenalectomy where cortisol has been markedly elevated sufficient to suppress by feedback inhibition of pituitary ACTH, hydrocortisone should be given. ACTH administration may help stimulate the contralateral adrenal, also.

***Addison's Disease** (**Primary Adrenal Cortical Insufficiency**)*

The various signs and symptoms demonstrated by the patient with Addison's disease are related to deficiencies occurring in the main adrenal cortical hormone group and to the increased secretion of pituitary ACTH and MSH (melanocyte-stimulating hormone). Low levels of cortisol produced in the deficient adrenal cortex results in the loss of the usual inhibition of ACTH production, so that ACTH increases above the normal.

In the clinical findings, hypotension is the consequence of absent or low levels of aldosterone. Also, the blood volume is reduced and serum sodium is greatly depressed. This latter results from the increased loss of sodium in the urine through the intestinal secretions and through perspiration. The addisonian patient frequently shows a mild to a moderate acidosis, since the sodium normally exchanges at the renal tubular level for potassium and hydrogen. With a lowered aldosterone secretion, also, potassium is retained. With the retention of potassium and hydrogen, the acidosis develops.

The lack of cortisol leads to weakness and anorexia. The patient shows poor cerebration. Characteristic of adrenal insufficiency, also, is an inadequate water diuresis. With the diminution of lack of secretion of 17-ketosteroids, the patient will show a loss of pubic hair, a loss of muscle mass, and weakness.

From the standpoint of fluids and electrolytes, the addisonian patient has a low blood volume, low sodium, bicarbonate, and chloride, and the potassium is elevated, creating a sodium-to-potassium ratio of less than 30.

In Addison's disease a deficiency of aldosterone causes a decrease in serum sodium with an increase in potassium. A determination of sodium alone may be misleading because of the hypohydration associated with Addison's disease. However, calculating the ratio between sodium and potassium will give a more reliable index of adrenal cortical function. Normally the ratio of sodium to potassium is 30 to 1. Any figure below this strongly suggests Addison's disease.

(The patient with secondary adrenal cortical insufficiency has some of the same clinical features of the addisonian patient, to include weakness, slow cerebration, loss of sex hair, and mild hypotension. However, these secondary changes are due to atrophy of the zona fasciculata and reticularis of the adrenal cortex. The zona glomerulosa, where aldosterone is formed, remains intact as it is independent of the trophic effect of ACTH. Therefore, aldosterone levels are normal and there is no hypohydration. In fact, sodium, bicarbonate, chloride, and potassium levels are normal and the sodium-potassium ratio is in the normal range of 30 or more.)

The response to anesthesia of the addisonian patient and for the patient with secondary adrenal cortical insufficiency will be the same, i.e. pronounced hypotension. Both require hydrocortisone for protection against this response, and the addisonian patient also requires replacement with sodium and chloride but not with potassium.

REFERENCES

1. Crout, J.R., and Brown, B.R.: Anesthetic management of pheochromocytoma: the value of phenoxybenzamine and methoxyflurane. *Anesthesiology, 1:*29, 1969.
2. Giesecke, A.H.; Spier, C.J.; Stanley, V.F., and Seltzer, H.S.: Considerations in the anesthetic management of the diabetic patient. *Clin Anesth, 3:*54, 1963.
3. Giesecke, A.H.; Spier, C.J., and Jenkins, M.T.: Management of diabetes mellitus during anesthesia and surgery. *Texas J Med, 60:*840, 1964.
4. Jenkins, M.T.; Giesecke, A.H., and Shires, G.T.: Electrolyte therapy in shock: management during anesthesia. *Clin Anesth, 2:*39, 1965.
5. Lamberth, I.E.: Obesity and anesthesia. *Clin Anesth, 3:*55, 1968.
6. Liddle, G.W.: Summarization of the effects of hormones on water and electrolyte metabolism. In Williams R.H. (Ed.): *Textbook of Endocrinology, 4th ed.* Philadelphia, Saunders, 1968.
7. Merrill, J.P.: *The Treatment of Renal Failure, 2nd ed.* New York, Grune & Stratton, 1965, Chap. 2.
8. Miller, W.F. and Bashour, F.A.: Cardiopulmonary changes in obesity. *Clin Anesth, 3:*127, 1963.
9. Netter, F.H.: *Endocrine System and Selected Metabolic Diseases.* Ciba Collections of Medical Illustrations, 1965, vol. 4, sections I and III.
10. Ploss, W.R., and Warner, W.A.: Pituitary surgery for diabetic retinopathy: problems in anesthetic management. *Clin Anesth, 3:*345, 1968.
11. Weisberg, H.F.: *Water, Electrolyte, and Acid-Base Balance,* 2nd ed., Baltimore, Williams & Wilkins, 1962, Chap. 2.
12. Welt, L.G.: *Clinical Disorders of Hydration and Acid-Base Equilibrium,* 2nd ed. Boston, Little Brown, 1959

Chapter 13

Comments on ADH

Tsutomu Oyama

Urine production is the result of three general processes: glomerular filtration, tubular reabsorption, and tubular secretion. The reabsorption of water is influenced by many factors, especially the solute load presented for excretion and level of posterior pituitary antidiuretic hormone (ADH, vasopressin) activity.

It is known that the oliguria that occurs during anesthesia and operation continues into the early postoperative phase and is influenced by alterations in renal function and osmolar clearance. The primary physiological functions of ADH are the regulation of the osmotic pressure of the extracellular fluid and the regulation of blood volume. However, no report has been made concerning the effect of anesthesia on human plasma ADH. For these reasons we studied seventeen patients (11 for ether, 6 for methoxyflurane) to measure directly peripheral plasma ADH levels by the modified bioassay method of Weinstein, Share, and Yoshida. Anesthesia alone was given for at least two hours before starting operation.

The measurements of osmolality in serum and urine were performed simultaneously as was urine output. Plasma ADH level was elevated remarkably to $16.9 \pm 3.5\mu U/ml$ (which was 6 times as much as pre-induction level) after thirty minutes of ether anesthesia alone. This was followed by gradual decrease in two hours after the start of anesthesia. Plasma ADH was elevated again to $34.5\mu U \pm 11.7\mu U/ml$ after the start of operation. Decrease in urinary output and increase in urinary osmolality were observed following at short intervals after the rise of plasma ADH.

It was noted that the lighter the level of ether anesthesia, the higher the plasma ADH level rose. On the contrary, the deeper the ether anesthesia, the lower the plasma ADH level fell.

The effect of methoxyflurane anesthesia alone on the plasma ADH level is less than that of the ether anesthesia. Blood volume is an important factor in the control of ADH release. Namely, the ADH is released in response to a depletion of blood volume. This might be physiologically important for the conservation of water so that blood volume can be restored. Other important releasing factors are the hypothalamus pituitary posterior axis, and other baroreceptors.

In operations of the extremities or orthopedic procedures, elevation in the plasma ADH level was less predominant than in laparotomy. The magnitude of this response is in proportion to the severity of the surgical procedure and the visceral traction. According to Moran, a gradual decline of ADH to normal values is usually observed by the fifth post-operative day in major cases.

The result of our study indicates that anesthesia and surgery induce decreased urine output not only by the impaired renal function but also by the increased plasma ADH concentration in the patient. However, methoxyflurane anesthesia showed significant but weaker effect on plasma ADH level than ether.

Chapter 14

Dynamics of Blood Glucose Regulation in Diabetic Patients and Normal Subjects Studied with Continuous Automated Analysis

GEORGE D. MOLNAR, F. JOHN SERVICE, JOHN W. ROSEVEAR, AND EUGENE ACKERMAN

During the past half-century, millions of lives have been saved or prolonged as a result of the discovery of insulin. Much new knowledge about diabetes has accumulated in that time, but progress toward a truly scientific basis for the day-to-day therapy of severe diabetes has been relatively slow. A major cause of this limited progress has been the lack of sufficient information about blood glucose (BG) dynamics in diabetics during their day-to-day existence, as compared to that of nondiabetics. Another major obstacle to progress has been the lack of sufficient information about insulin levels in the circulation of diabetics, especially in those who are treated with insulin.

The 1960's provided two relevant technical advances. One was the development in this country[1] and in France[2] of continuous automated blood glucose monitoring (CBGM). The other was the development of accurate assays for insulin in the circulation. This chapter describes some of our experiences with CBGM. Chapter 16 will describe testing of insulin secretory reserve of insulin-treated patients and measuring the levels and patterns of circulating plasma insulin resulting from therapy with insulin, subjects which have been of special interest to us and our colleagues.[3]

When CBGM and immunoreactive insulin (IRI) measurements are combined, and even additional data are accumulated simultaneously, it becomes feasible to assess in depth the adequacy and limitations of current therapy for diabetes. Only from such an assessment can we progress to better treatment for our patients under all conditions, including before, during, and after anesthesia. The same assessment in depth also provides the clues for needed improvements of current therapeutic agents and regimens.

This chapter considers how optimally and suboptimally treated diabetics compare to one another on comparable therapeutic regimens and how diabetics compare to nondiabetics as well. We are working toward definitions and quantifications of normalcy. By comparing diabetics with normals, we propose to define and to quantify the degree to which we have been able to imitate nature in our therapy. In this sense, perfect control of diabetes is complete normalization of the diabetic patient through therapy. To find out whether this is feasible and how to approach such lofty goals are challenging problems in diabetic therapeutics.

This topic has some relevance to anesthesiology. Protection from excessive hypoglycemia or hyperglycemia before, during, and after anesthesia is of great importance for the diabetic patient. In addition, from the methods and data presented here, it may be possible to extrapolate to other situations, such as BG and IRI behavior during operation for insulin-producing islet cell tumors or behavior of BG, IRI, and other hormones during open-heart surgery, with special attention to the effects of hypothermia.

METHODS

Table 14-I describes the six normal subjects we studied, three men and three women. All were within 15 per cent of ideal weight, and their freedom from diabetes was demonstrated by tests with glucose, tolbutamide, and insulin.

Table 14-I also describes the twelve diabetics on whom we have completed studies so far, four men and eight women. The age range was twenty-one to seventy years. Body weights were within 15 per cent of the ideal level. The duration of the diabetes was nine to forty-one years. The therapy was insulin except in patients D9 and D12, who were on oral hypoglycemic agents, but they too had once been treated with insulin. In regard to the degree to which clinical management was difficult or easy, on the basis of clinical assessment the patients were classified as follows: high degree of unstable diabetes, three; moderately unstable diabetes, five; and stable diabetes, four.

The deliberate preponderance of unstable diabetics among our patients made it unlikely that the usual intermittent methods of blood-sampling would describe their BG fluctuations adequately. We clearly needed CBGM. Moreover, we were interested in clinical phenomena such as hypoglycemia and muscular exercise. These occur under conditions such that we were obliged to study ambulatory, fed subjects over extended periods. We needed standardized living conditions, to permit comparisons in the same subject from time to time as well as between subjects. Furthermore, we needed simultaneously obtained data on other variables in addition to BG changes.

The BG monitoring required the development of a method for continuous *in vivo* analysis. The advantages of Dr. Rosevear and associates'[5] method over previous methods include specificity (it is a glucose oxidase

TABLE 14–I
RELEVANT CLINICAL DATA FOR SUBJECTS STUDIED

Subject	*Age (yr.); Sex*	*Height (in.)*	*Weight (lb.)**	*Diabetes Duration (yr.)*	*Usual Therapy†*	*Degree of Stability‡*	*Complications or Associated Conditions§*
NONDIABETIC (normal)							
N1	24 M	72½	184 (180)				
N2	57 F	60½	120 (115)				
N3	24 F	64¼	112 (120)				
N4	32 M	67	180 (165)				
N5	23 M	74½	177 (180)				
N6	28 F	68¾	127 (135)				
DIABETIC							
D1	45 F	65½	120 (120)	41	N + R, 2	HU	NLD (MS); PN (M)
D2	44 F	64¾	111 (125)	12	N + R, 2	HU	Ischemic heart disease
D3	20 F	66	124 (125)	15	SL + L, 1	HU	Retinopathy (MS); PN (M)
D4	59 F	64	143 (140)	20	R + SL + UL, 1	U	Retinopathy (M); PN (M)
D5	35 M	66½	145 (145)	33	R + L, 1	U	None
D6	36 M	69¾	162 (160)	24	R, 3; SL, 1	U	Retinopathy (MS); nephropathy (M)
D7	21 M	69	158 (155)	12	L, 2	U	NLD
D8	25 F	63	134 (120)	9	L + U, 1	U	Retinopathy (M)
D9	47 F	62½	113 (116)	11	T, 1; P, 2	S	Insulin allergy; rheumatoid arthritis, inactive
D10	55 M	68	156 (155)	14	L, 1	S	Retinopathy (MS); PN (M)
D11	70 F	64	175 (140)	18	L, 1	S	Retinopathy (S); PN (M); ischemic heart disease
D12	59 F	67½	161 (150)	10	D, 1; P, 1	S	Acromegaly

From Molnar and associates.[4]
* Ideal weight is shown in parentheses.
† Numbers indicate frequency of insulin injections or number of doses of oral agent per day; + indicates insulins are given in same syringe; N = NPH, L = lente, SL = semilente, UL = ultralente, R = regular, T = tolazamide (Tolinase®), P = phenformin, D = acetohexamide.
‡ HU = highly unstable, U = unstable, S = stable.
§ NLD = necrobiosis lipoidica diabeticorum, PN = peripheral neuropathy; severity of complications: M = mild, MS = moderately severe, S = severe.

method), speed of response, and width of range[5] (range is 0 to 800 mg/100 ml of blood). Sensitivity can be maintained for twenty-four hours or longer and then restored by rinsing the analytical system, permitting studies of more than forty-eight hours' duration. The amount of blood which is continuously withdrawn, from the systemically heparinized patient, for continuous analysis is only 6 ml/hr. This small volume even provides blood for additional measurements, e.g. radioimmunoassay of insulin or growth hormone (HGH). To accomplish these additional measurements, at the time of introducing the standard glucose solution into the sampling line to calibrate the instrument (done once an hour), the blood is collected separately instead of being allowed to traverse the analytical system.

The patient is attached to the analytical equipment (Fig. 14-1) by means of an intravenous Teflon® catheter, long enough (10 feet) to permit freedom of movement. The catheter extends from the left antecubital fossa to the modified AutoAnalyzer® unit mounted on a mobile cart.

The general plan of the investigation included the following. During

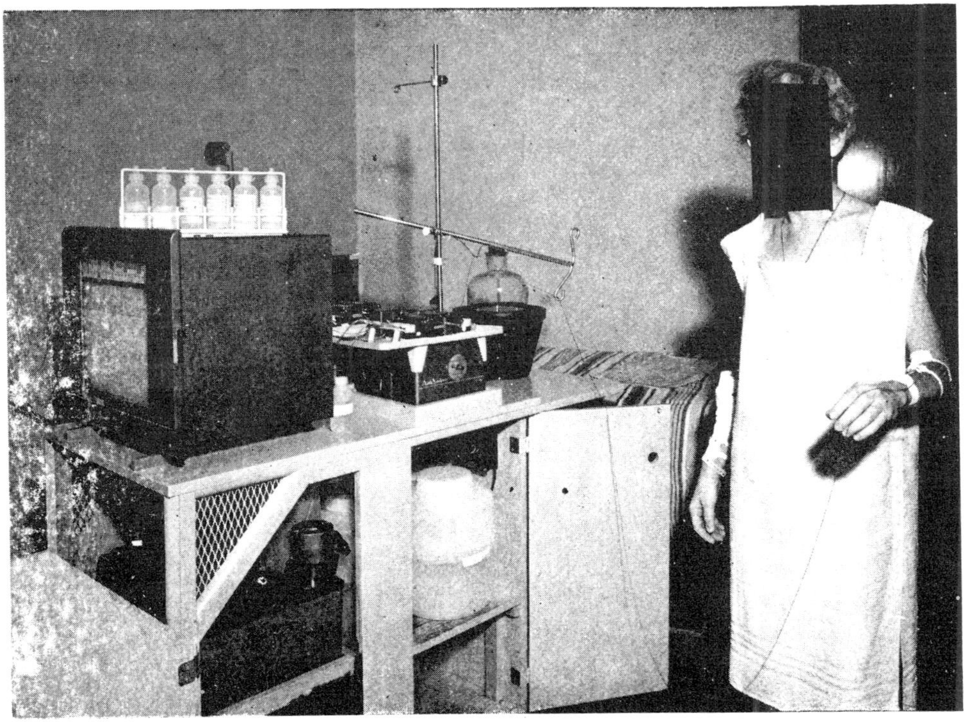

Figure 14-1. Mobile analyzing unit. (From Rosevear, J.W.; Pfaff, K.J.; Service, F.J.; Molnar, G.D., and Ackerman, E.[5] By permission of Hoeber Medical Division, Harper & Row, Publishers.)

TABLE 14-II
SUMMARY OF ANALYTICAL METHODS

Variable	*Specimen Analyzed*	*Method*
Glucose	Blood	Glucose oxidase
	Urine	Ferricyanide
Ketone bodies	Plasma	Distillation and trapping
	Urine	
Free fatty acids	Plasma	Extraction and titration
Growth hormone	Plasma	Radioimmunoassay
Insulin	Plasma	Radioimmunoassay
Nitrogen	Urine	
	Stool	Kjeldahl
	Diet	

From Molnar and associates.[4]

CBGM, additional specimens of blood for the determinations listed in Table 14-II were obtained from another catheter fitted with a stopcock and usually placed in a forearm vein. For these tests, urine was collected in periods lasting from one hour before each meal to one hour prior to the next meal. Collections of stool were made during the six-day balance periods.

A six-day metabolic balance technique was used before, during, and after the forty-eight-hour long CBGM. In this manner, the more intensive forty-eight-hour portion of the study was extended by ninety-six more hours of less demanding study.[4]

Before and after the six-day balance study the patients lived in a metabolic ward. This permitted diabetic regulation to an optimal level in preparation for the balance study. The shorter tests, to measure responses of insulin and growth hormone secretion, under resting and fasting conditions, to glucose, tolbutamide, isulin, glucagon, arginine, epinephrine, and saline were done at these times.[4]

Each subject's activities throughout his stay in the metabolic ward were regulated according to the schedule shown in Table 14-III. For

TABLE 14-III
DAILY SCHEDULE OF ACTIVITIES

Activity	*Time (24-hr clock time)*
Insulin injection*	7:45; 12:45; 17:45; 22:45
Meals	8:00; 13:00; 16:00†; 18:00; 23:00
Exercise	10:00–11:00; 15:00–16:00; 20:00–21:00
Blood sampling	7:45; 11:00; 16:00
Urine:	
Voidings‡	7:00; 7:30; 12:00; 12:30; 17:00; 17:30; 22:00; 22:30
Collections§	7:00–12:00; 12:00–17:00; 17:00–22:00; 22:00–7:00

From Molnar and associates.[4]
* If once daily, at 7:45; if twice daily, at 7:45 and 17:45.
† Snack.
‡ Specimens for semiquantitative tests by nurses.
§ Total collection for quantitative analysis.

convenience of notation, twenty-four-hour clock time was used in recording all events of the day or night. Insulin injections were given fifteen minutes before meals, meals were given at five-hour intervals, and exercise was performed in the third postprandial hour. Once a patient's optimal diet therapy was worked out, it was kept constant within and between days. To provide data on constancy of BG and other variables, the major meals were designed to have equal composition within as well as between twenty-four-hour cycles. This applied to calories as well as to carbohydrate, fat, and protein. A small afternoon snack was also included because this is customary in the treatment of severe diabetics.[4]

Exercise was standardized as walking exercise, whether or not the subject was attached to the analytical equipment. The prolonged preparation of the patients trained them well for walking at a constant pace, for one hour three times daily, in all phases of the investigation.

Insulin therapy is, without a doubt, the most potent diabetic therapeutic factor. This was adjusted, while diet and exercise were kept constant, to provide such further improvement of diabetic management as was feasible.

Each unstable diabetic was studied at least once on a regimen of four daily injections of short-acting insulin and at least once on a less intensive regimen of longer-acting insulin injected once or twice daily. Each insulin regimen was worked out uniquely, for optimal management of the patient, during the preparatory studies.

QUALITATIVE RESULTS

Because of the large number of variables and the mass of data, a computer program was developed to assist in the necessary calculations, recording, and even plotting of accurate graphic summaries of each experiment. Such a graphic summary of data is shown in Figure 14–2. It illustrates, in the case of a normal subject, the schedule of activities across the top with the simultaneous multiple measurements graphed below it, all coordinated with the forty-eight-hour time scale on the abscissa. Note the rhythmic undulations of the continuous BG record within the range of 60 to 140 mg/100 ml. Each meal is associated with an increase and a return to the base line.

Figure 14–3 is another computer-drawn graphic summary with typical results from an unstable diabetic patient; in this case the events listed across the top include an injection of insulin once a day at 7:45 A.M. Note the wide, erratic fluctuations of the BG, to a peak as high as 450 mg/100 ml at one point and to hypoglycemia at another.

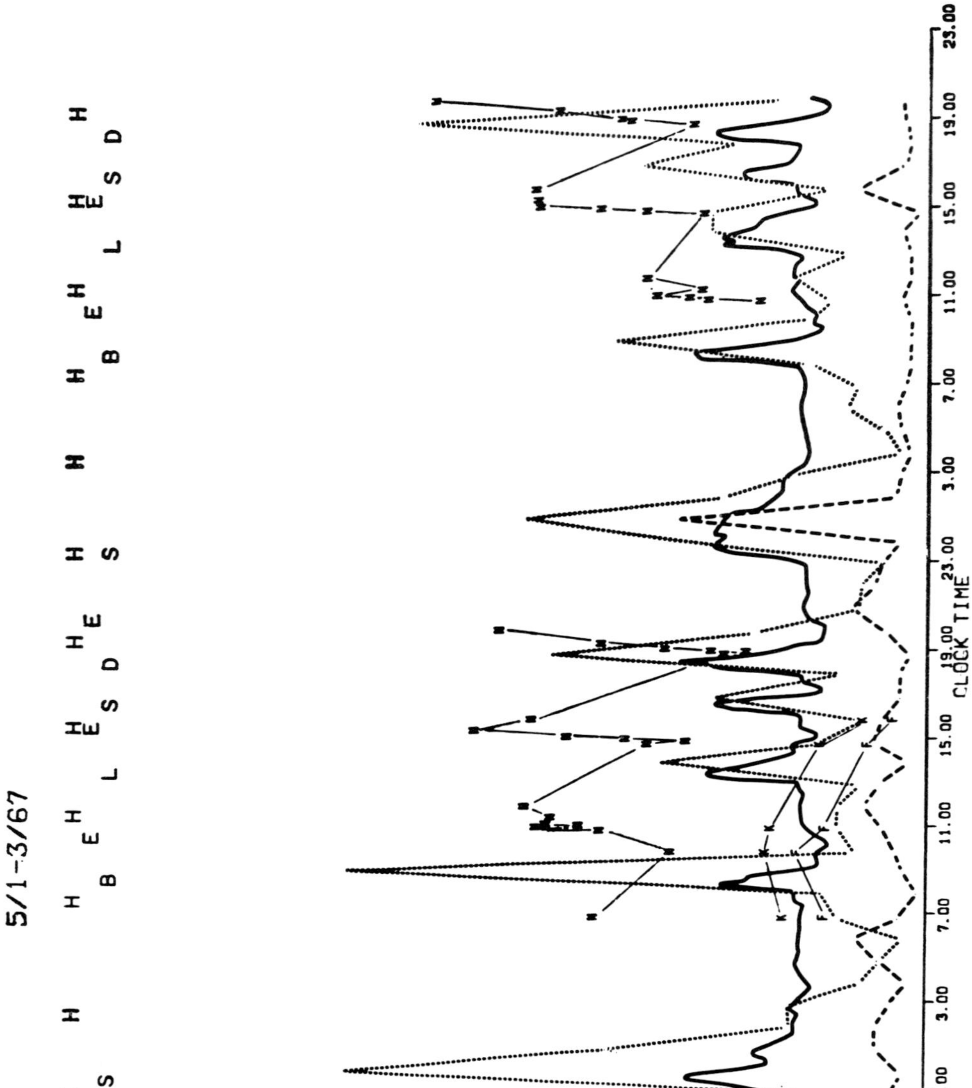

Figure 14–2. Summary of data from normal subject (N1) during forty-eight-hour CBGM with simultaneous measurements of variables in plasma. Illustration is computer-drawn composite with some lines reinforced for clarity. BG concentration = solid lines; HGH = dashed line; IRI = dotted line; N = NEFA; K = total KB; F = free KB; H = heparin injection; E = exercise; B = breakfast; L = lunch; S = afternoon snack at 16:00 or supper at 23:00; D = dinner. (From Molnar, G.D., Ackerman, E., Rosevear, J.W., Gatewood, L.C., and Moxness, Karen E.[4] By permission.)

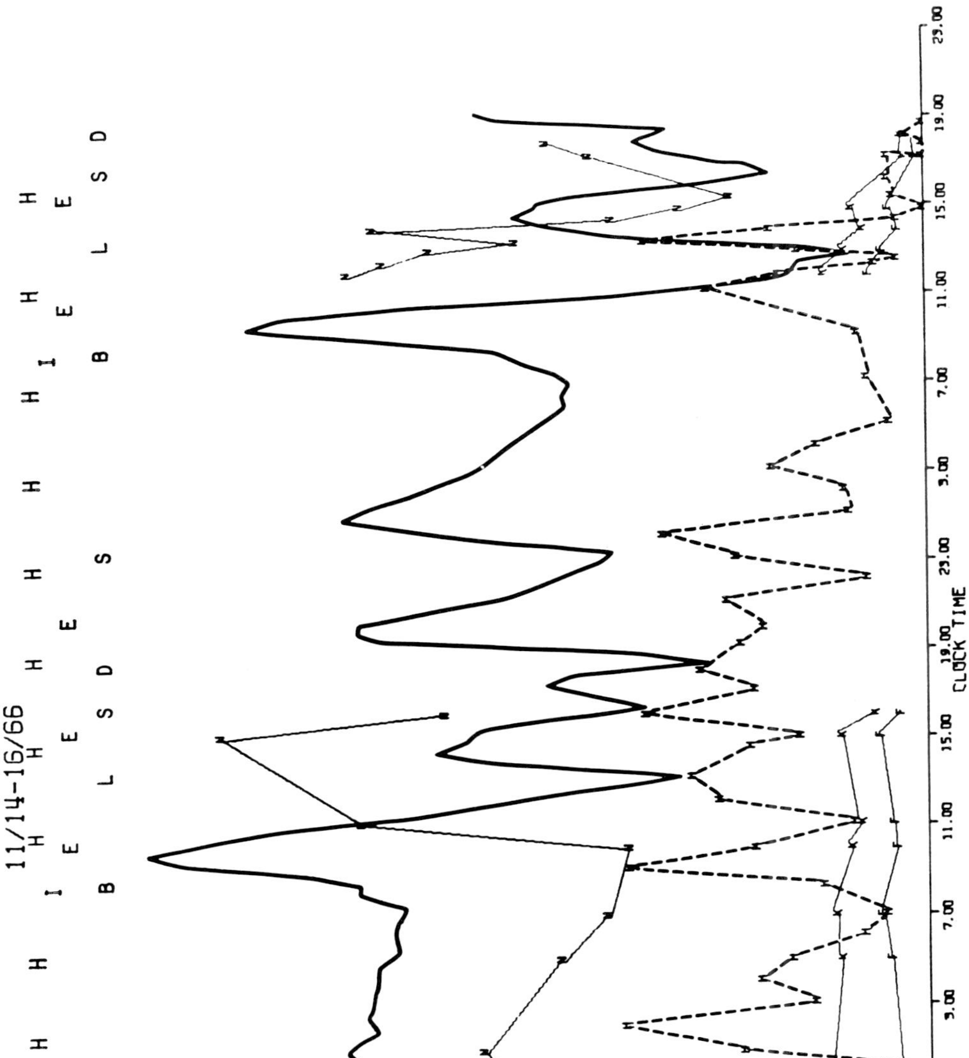

Figure 14–3. Summary of data from unstable diabetic subject (D2) during a forty-eight-hour CBGM with simultaneous measurement of variables in plasma. For explanation see Figure 14–2. I = injections of insulin at 7:45. (From Molnar, G.D., Ackerman, E., Risevear, J.W., Gatewood, L.C., and Moxness, Karen E.[4] By permission.

We found it necessary to use systemic anticoagulation with heparin during CBGM. Because of the increased free fatty acid levels resulting from the effect of heparin on lipoprotein-lipase, we performed studies under matching conditions without use of heparin.[6] Figure 14–4 illustrates such studies in a normal, a stable diabetic, and an unstable diabetic. These studies compared matched BG determinations from the continuous record obtained during systemic heparin administration with those made on intermittently obtained blood specimens from an indwelling catheter but without heparin therapy. Note the similarities of pattern within the same individual. The contrasts between the normal, the stable diabetic, and the unstable diabetic are also interesting. Notice the widening swings of the BG. Intensive statistical evaluation showed that the heparin had no significant effect on BG levels or carbohydrate tolerance, despite increased free fatty acid levels.

To focus once again on the combination of CBGM, which of necessity is limited to forty-eight hours' duration, with such metabolic investigative methods as can be done over longer periods, we turn to the composite charts shown in the next two figures. One of the objectives of these studies was to provide data over more extended periods. Another was to evaluate how representative the forty-eight-hour CBGM was of the longer periods. By correlative analysis of the urinary glucose excretion during two days of CBGM with that during the four other days of the balance period, the CBGM was found to be representative of the longer periods.

Figure 14–5, a composite of portions of two studies on the same patient, combines time-coordinated graphic patterns which represent insulin dosage; insulin reactions; urinary glucose, ketone body, and nitrogen excretion; BG and plasma HGH levels, and nitrogen balance. The days in November 1966 were during a therapeutic regimen of a single daily injection of insulin (which happened to be a mixture of three insulins). There were many insulin-induced hypoglycemic reactions. The urinary glucose swings were wide and irregular, as were the BG swings. The ketone body, HGH, and nitrogen balance data all show some consistent patterns. However, the main point is the location, with respect to time, of the CBGM in the middle of the balance period. The patient was studied again, during a therapeutic regimen of four daily injections of short-acting insulin (note that a tolbutamide tolerance test was performed). Diabetic regulation seems to have improved and there are fewer hypoglycemic reactions, less glycosuria, and a more rhythmic BG pattern. Again, note the location of the CBGM during the middle of the balance period.

Figure 14–6 shows segments of two studies in a stable diabetic sub-

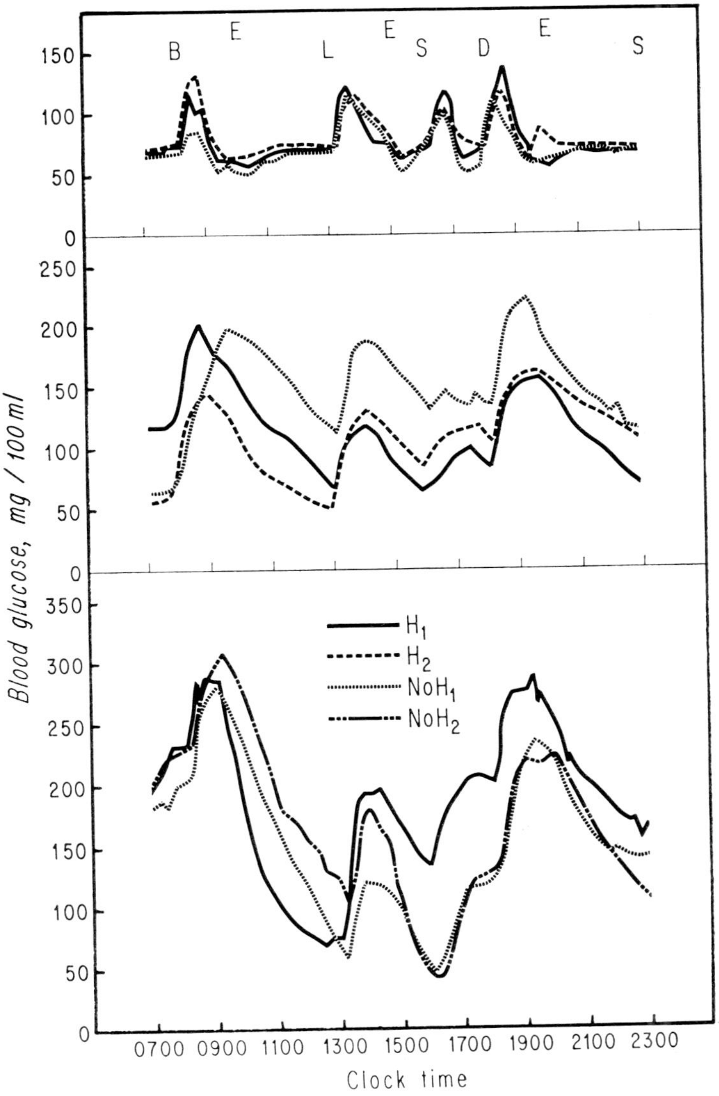

Figure 14–4. Diurnal BG patterns during CBGM with heparin administration and BG analysis on intermittently obtained blood specimens without heparin administration. Symbols for events (*across top*): B = breakfast; E = 1 hour's exercise; L = lunch; S = afternoon snack at 16:00 and supper at 23:00 D = dinner. Key to lines: H_1 and H_2 = first and second days with heparin administration; NoH_1 and NoH_2 = first and second days without heparin administration. *Top,* Normal subject. *Middle,* Stable diabetic. *Bottom,* Unstable diabetic. (From Service, F.J., Molnar, G.D., Rosevear, J.W., Ackerman, E., Taylor, W.F., Cremer, G.M., and Moxness, Karen E.[6] By permission.)

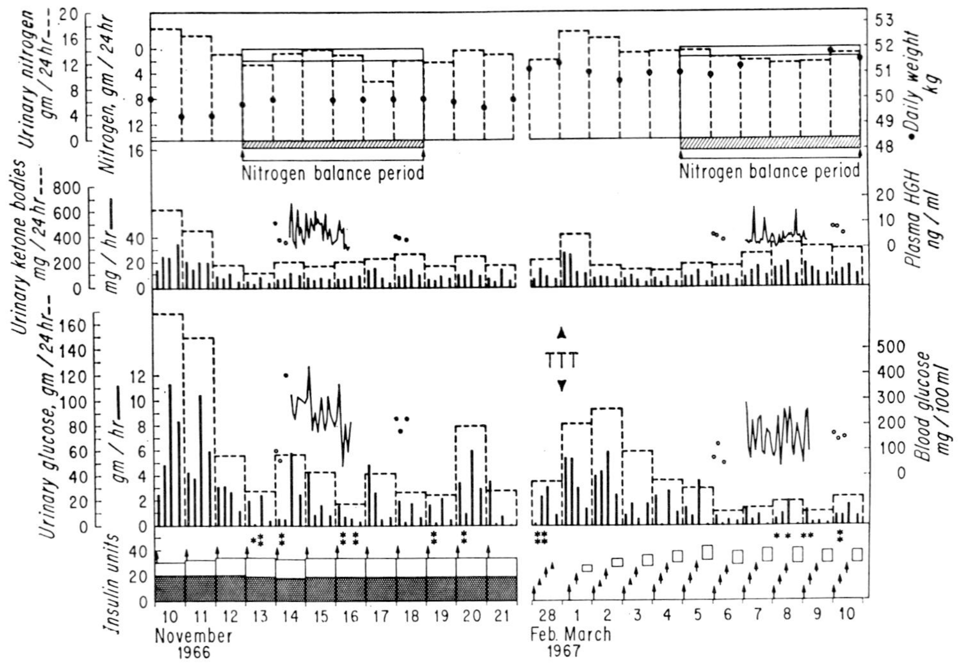

Figure 14–5. Composite metabolic chart for unstable diabetic (D2). Patient was on regimen L for first part and on regimen S for second. Within each twenty-four-hour period (from 7:00 to 7:00), insulin injections, hypoglycemic episodes, values for urinary glucose and KB excretion and for BG glucose and plasma HGH are indicated at the hour of occurrence or collection. Tolbutamide tolerance test is indicated by TTT. Insulin therapy: arrow = regular insulin; crosshatched area = ultralente insulin; open area = semilente insulin. Total daily dose is indicated cumulatively upward. Symbols: asterisk = mild hypoglycemic episode; double asterisk = moderately severe hypoglycemic episode. Solid vertical bars = glucose of KB excreted per hour in the four segments of each day. Bar defined by broken lines = total glucose of KB or nitrogen excreted per twenty-four hours. Blood glucose and HGH = continuous line showing hourly pattern of values during CBGM (*scales at right side*). Open circles indicate intermittently determined values. Daily weight is indicated by solid circles. Nitrogen balance is conventionally charted (positive balance indicated by a clear zone below zero line); crosshatched zone represents fecal excretion of nitrogen. (From Molnar, G.D., Ackerman, E., Rosevear, J.W., Gatewood, L.C., and Moxness, Karen E.[4] By permission.)

ject. All is much calmer on this graph. There is only one hypoglycemic episode. The urinary glucose and BG patterns are low and relatively regular. Note again that the CBGM occurred in the middle of the balance period.

Although the qualitative differences are evident between individual BG curves and between groups of BG curves, such as between normals and diabetics and between stable and unstable diabetics, the value of

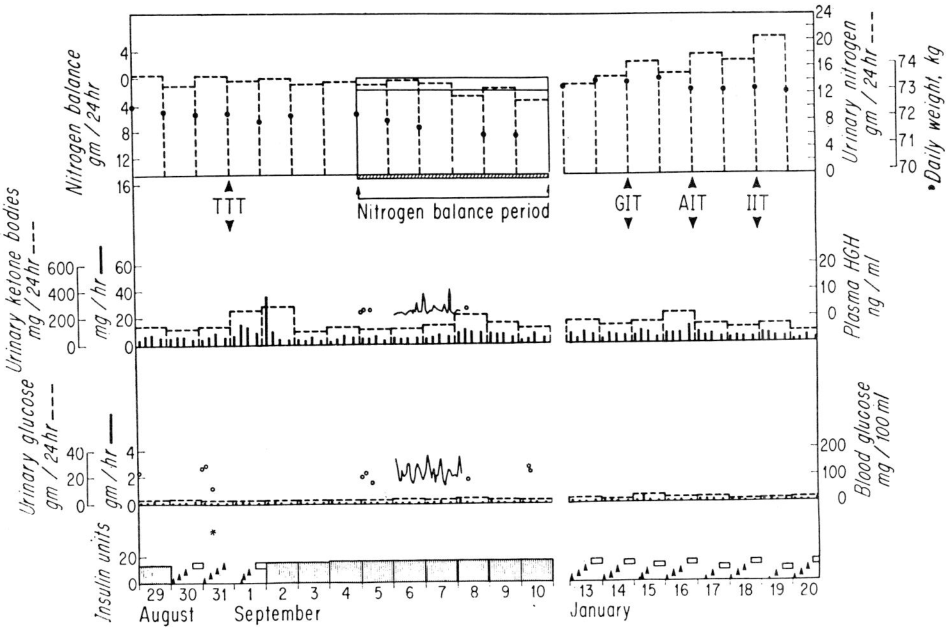

Figure 14–6. Record as in Figure 14–5, for stable diabetic patient (D10). GIT = glucagon infusion test; AIT = arginine infusion test; IIT = insulin infusion test. (From Molnar, G.D., Ackerman, E., Rosevear, G.W., Gatewood, L.C., and Moxness, Karen E.[4] By permission.)

such qualitative analyses is limited. We, therefore, sought to quantitate the differences.

QUANTITATIVE RESULTS

One quantitation involved the determination of the mean diurnal level of BG (MBG). This was accomplished by computing the arithmetic average of 288 BG values taken from the continuous record, representing five-minute intervals during twenty-four hours. Examples of these MBG values are shown in Table 14-IV. Note that there is no overlap in values among the three groups. This implies that, with the best of efforts with a given program of insulin treatment, we were unable to reduce the MBG in unstable diabetics to the level of the stable diabetics. Similarly, we failed to reduce the MBG of stable diabetics to the level of normal subjects.

Another quantitation involved determination of the mean amplitude of glycemic excursions (MAGE) during the forty-eight-hour CBGM period. This calculation uses the arithmetic average of BG increases from nadir to peak or BG decreases from peak to nadir. We found that in normal subjects the postprandial BG increases and decreases exceeded

TABLE 14–IV
MEAN BLOOD GLUCOSE (MBG) VALUES

Subject	*Mg/100 ml*
NORMAL	
N1	79.7
N2	82.7
N3	78.1
STABLE DIABETICS	
D11	110.7
D12	110.1
D9	110.0
D10	114.6
UNSTABLE DIABETICS	
D4	195.8
D6	170.5
D7	146.2
D8	149.2
D1	207.8
D2	244.1

one standard deviation of the MBG. We also found that diabetics sometimes had BG oscillations of questionable significance. Therefore, a BG increase or decrease had to exceed one standard deviation (of the MBG) to be a glycemic excursion.

The MAGE in a representative sample of our studies is shown in Table 14-V. It is evident that unstable diabetics have larger MAGE than do stable diabetics, and stable diabetics, in general have larger MAGE than do normal subjects. What this means is that, in diabetics, with the best of our efforts on a given program of treatment, we were unable to reduce, below a certain level, the average BG swings resulting from meals and exercise, by therapy with insulin. It further means that, with higher-than-normal MAGE, the BG increases too much after meals, despite the insulin we gave, and eventually decreases too much because of the insulin we gave and because of the physical activity. The lowest

TABLE 14–V
MEAN AMPLITUDE OF GLYCEMIC EXCURSIONS (MAGE)

Subject	*Mg/100 ml*
NORMAL	
N1	59.7
N2	44.4
N3	22.5
STABLE DIABETICS	
D11	67.4
D12	55.9
D9	81.7
D10	78.6
UNSTABLE DIABETICS	
D4	154.4
D6	169.1
D7	119.0
D8	148.8
D1	200.0
D2	183.9

individual level of MAGE achieved with the most intensive therapeutic efforts, in a series of studies of each patient, has so far been the most precise way of expressing the degree of given patient's diabetic instability.[7] Normal subjects serve as the standard of stability. Some stable diabetics approach that standard. Unstable diabetics do not do so.

To explore further the meaning of this measure of diabetic instability, the following studies and calculations were carried out. Patients were treated with one daily injection of insulin on our standard program of diet and exercise.[4] Then they were treated with an intensified insulin regimen using short-acting insulin in four daily injections. Although the diurnal MBG could at times be reduced by nearly 100 mg/100 ml, the extent to which the MAGE could be decreased was always much less. Table 14-VI demonstrates that, although intensified therapy decreased the MBG and the urinary glucose excretion significantly, the MAGE did not change significantly.

TABLE 14–VI
RESULTS OF PROGRAMS OF LONG-ACTING INSULIN (L) AND SHORT-ACTING INSULIN (S) THERAPY IN UNSTABLE DIABETICS

	MBG (mg/100 ml)		*MAGE (mg/100 ml)*		*99.7% Upper Confidence Limit of Urinary Glucose (gm/hr)*	
Subjects	*L*	*S*	*L*	*S*	*L*	*S*
D4	195.8	113.3	154.4	131.3	5.60	1.20
D6	170.5	101.1	169.1	123.4	1.50	0.34
D7	146.2	140.9	119.0	133.7	6.32	3.80
D8	149.2	118.8	148.8	118.9	5.12	0.55
D1	207.8	144.8	200.0	197.8	3.52	3.60
D2	244.1	152.0	183.9	184.7	6.15	1.97
Mean diff.	+56.82		+14.13		+2.79	
SD	32.75		22.38		1.93	
P	<0.005		>0.05		<0.01	

INTERPRETATIONS

Intensification of insulin therapy has its benefits and its limitations. The benefits are that, with dietary therapy and exercise already being optimal, the MBG may be improved and the urinary wastage of glucose, ketone bodies, and nitrogen may be decreased. The patient's sense of well-being improves. The limitations are that, because the MAGE cannot be further decreased, the patient's individual BG pattern of early postprandial increases and late postprandial decreases undergoes an overall downward shift. Such a downward shift results in hypoglycemic episodes occurring near the BG nadirs. With too many such hypoglycemic episodes and at BG nadirs which are too low, the intensified therapy can result in an intolerable number of insulin reactions of excessive severity. A middle ground must obviously be accepted as a compromise. It is the MAGE of a given individual, on a given therapeutic regimen, which determines the extent to which that patient's BG regulation may be improved by therapy.[7] It is in this sense that MAGE is a measure of that patient's diabetic instability and a criterion of the true difficulty of optimal diabetic management.

SUMMARY

Continuous monitoring of blood glucose, in ambulatory fed subjects, with simultaneous frequent intermittent determinations of other variables was combined with prolonged metabolic investigation during a standardized scheme of therapy and activities. Subjects included stable and unstable diabetics and normals. During attempts at optimal diabetic therapy, diurnal mean blood glucose (MBG) levels and mean amplitude of glycemic excursions (MAGE) distinguished among these three groups. With intensified therapy with insulin, MBG could be decreased but MAGE could not.

REFERENCES

1. Weller, C.; Linder, M.; Macaulay, Alice; Ferrari, A., and Kessler, G.: Continuous *in vivo* determination of blood glucose in human subjects. *Ann NY Acad Sci, 87*:658–668, 1960.
2. Mirouze, J.; Jaffiol, C., and Sany, C.: Enregistrement glycemique nycthemeral continu dans le diabete instable. *Rev Fr Endocrinol Clin, 3*:337–353, 1962.
3. Cremer, G.M.; Molnar, G.D.; Taylor, W.F.; Service, F.J.; Gatewood, Lael C.; Ackerman, E., and Rosevear, J.W.: Studies of diabetic instability. II. Tests of insulinogenic reserve with arginine, glucagon, epinephrine, and saline. *Metabolism, 20*:1083–1093, 1971.
4. Molnar, G.D.; Ackerman, E.; Rosevear, J.W.; Gatewood, L.C., and Moxness, Karen E.: Continuous blood glucose analysis in ambulatory fed subjects. I. General methodology. *Mayo Clin Proc, 43*:833–851, 1968.
5. Rosevear, J.W.; Pfaff, K.J.; Service, F.J.; Molnar, G.D., and Ackerman, E.: Glucose oxidase method for continuous automated blood glucose determination. *Clin Chem, 15*:680–698, 1969.
6. Service, F.J.; Molnar, G.D.; Rosevear, J.W.; Ackerman, E.; Taylor, W.F.; Cremer, G.M., and Moxness, Karen E.: Continuous blood glucose analysis in ambulatory fed subjects. II. Effects of anticoagulation with heparin. *Mayo Clin Proc, 44*:466–477, 1969.
7. Service, F.J.; Molnar, G.D.; Rosevear, J.W.; Ackerman, E.; Gatewood, L.C., and Taylor, W.F.: Mean amplitude of glycemic excursions, a measure of diabetic instability. *Diabetes, 19*:644–655, 1970.

Chapter 15

Some Considerations in the Management of Nodular Goiter

BROWN M. DOBYNS

The term "nodular goiter" means different things to different persons. Actually, it is a very useful term when one is carrying out a physical examination and wishes to describe lumps in the thyroid gland. Under such circumstances, those lumps might represent a variety of pathologic conditions; such as adenoma, carcinoma, cysts, thyroiditis, or perhaps just a diffusely lobulated gland, which represents an hypertrophy of the thyroid which is a compensatory effort of the gland to make up for a deficiency of thyroid hormone. As long as we are palpating from the outside of the neck, the term "nodular goiter" is appropriate, because we cannot be certain what the lumps actually are. Once the tissue is available for even gross inspection and more particularly for microscopic study, the term "nodular goiter" is no longer appropriate. Cysts and thyroiditis are easily recognized when the tissue is available; but, unfortunately, many pathologists are reluctant to draw a distinction between discrete adenomas, on one hand, and the diffusely lobulated glands bearing varying areas of hyperplasia and involution, on the other. Both are commonly lumped together and called nodular goiter, even as a final pathological diagnosis.

Radioautographs have taught us to distinguish between true adenomas and the diffuse lobulated gland that sometimes appears to simulate adenomas. Some masses of cells in the thyroid, even though they are not clearly surrounded by a conspicuous capsule, have a distinctly different functional capacity than the remaining extranodular tissue. Radioautographs have shown this in a diffusely lobulated gland in which there may be a suggestion of connective tissue capsules. There are often no areas whose function is significantly different to demarcate these from other areas. The entire thyroid has responded to a demand to make more hormone. As that demand waxes and wanes, areas of involution and hyperplasia develop. The radioautograph shows a mottled distribution of radioactive iodine corresponding with the variations in hyperplasia, but no masses of cells stand out as being functionally different from other areas. In contrast, the radioautographs of the adenomatous glands show discrete areas composed of cells which are functionally different from the cells which make up the extranodular tissue. This is not to deny that discrete adenomas sometime occur in the diffusely lobulated gland, but it does show that adenomas when present are clearly distinguishable on the basis of the physiological function. The diffusely lobulated gland is interpreted to represent a chronic deficiency in thyroid hormone output, while the adenomatous gland often contains extranodular tissue which morphologically and functionally bears no evidence of present or past deficiency in hormone production.

Experience has now shown that when a subtotal thyroidectomy is performed upon the diffusely lobulated gland and its total volume reduced, the surviving remnant regenerates, resulting in recurrent "nodular goiter," because the removal of some tissue from a gland, which was already incapable of doing the job that it was expected to do, is made even less capable. On the other hand, in the other type of nodular goiter, experience has shown us that removal of discrete adenomas without sacrifice of the normal extranodular tissue seldom results in recurrence of the goiter and is not followed by the need for thyroid hormone supplement.

The clinicians' problem is one of trying to distinguish which so-called nodular goiters should be referred to a surgeon and which should not. If it is the diffuse lobulated gland representing a deficiency, supplemental hormone should be given. If there are palpable discrete masses found by vigorous aggressive palpation, there is no point in giving hormone, as will be shown later. The discrete masses should be removed. Sometimes the physician cannot be sure that the irregularities are discretely different from other areas of the thyroid, because the questionable masses are buried deep within a large boggy gland. He may then try the administration of thyroid hormone. If he elects to do this, he has assumed two responsibilities: (1) to give enough thyroid hormone to cause suppression of the physiologically responsive part of the gland (the nonadenomatous part) and (2) to make sure that he reobserves the gland in five or six weeks to see what is left after the normal part of the gland shrinks. If discrete masses are then detectable, they should be removed. Some internists have been lulled into complacency by giving thyroid and, although they have recognized some shrinkage of the goiter, they, as a result of a poor physical examination, have not distinguished between discrete masses which remain and the shrinkage of the extranodular tissue. By giving reassurance to the patient, they may have committed an error by overlooking a mass that may be a carcinoma.

There has developed in the minds of some physicians the idea that adenomas will shrink when supplemental thyroid hormone is given. Our studies have shown that this is by and large a false interpretation on the part of those who have had a chance only to palpate and have not had an opportunity to study the pathologic physiology.

As part of a special study in progress for a good many years, we have removed lesions after various lengths of time following their discovery and after a sequence of observations have been made on such masses with and without the administration of thyroid hormone. The following sequence of events becomes clear. Masses are initially rather insignificant adenomas. Hemorrhage or necrosis takes place within them. Owing to the decomposed tissue and the build up of osmotic pressure within, secondary swelling occurs. It is this event which is usually painless and which brings the patient to the doctor. The doctor, believing that thyroid hormone will make such lesions go away, promptly starts the administration of thyroid hormone. Actually, it is the normal sequence of events which results in the resolution of such a mass that makes the physician feel that the administration of the hormone was responsible for the change. Reasonably fresh hemorrhage may be found in some if they are removed the same day or the day after they appear. If one waits for a few weeks, chocolate brown necrotic tissue is often found. If one waits several weeks or months, the center of the necrotic area is reabsorbed and only a surviving rind of functioning thyroid tissue representing the adenoma remains inside the capsule. If one waits many months or years, the central scar contracts and regeneration of the scraps of tissue that represented the original tumor give rise to a cluster of daughter tumors. Each of the cluster, gives the same radioautograph, showing that the cells came from the same origin. The multiple tumors that recur are bound together by connective tissue that represented the original necrotic core. Thus, some patients are given thyroid hormone supplement and the physician does not see the patient again for that complaint. If the mass recurs conspicuously or continues to grow without apparent benefit from the medication, the patient tries another doctor. Unfortunately, some of these masses are malignant and there is undesirable delay in their removal.

The problem that confronts the surgeon is which of the "nodular goiters" require exploration and which do not. Unfortunately today there are surgeons operating on diffusely lobulated glands which should not be explored. Those surgeons are doing so because they have been startled to find a carcinoma in some of these glands when they really did not expect it. Radioactive iodine scanning may shed considerable light on the functional nature of discrete masses and show that some which correspond to the physical examination either have excessive or minimal radioactive iodine content. Those masses which are cold should be removed promptly.

Among the patients suffering from thyrotoxicosis, there are some who have Graves' disease and no discrete masses to be palpated. There are others who have had Graves' disease for quite some time and whose gland has become finely pebbly, but do not contain discrete masses. This is merely a result of involution and hyperplasia and indicates the disease is of considerable duration. This is a variation of the diffuse hyperplastic thyroid of Graves' disease and should not be called nodular goiter. There are also patients who are suffering from thyrotoxicosis which is attributable to hyperfunctioning single or multiple adenomas. Radioautographs prepared on more than 3,000 thyroid lumps over the past twenty years have shown us that of all the hyperfunctioning* adenomas found, only one fourth of those patients are suffering from clinical thyrotoxicosis at the time the lumps are removed. We knew from the observations of Plummer long ago that many years elapse between the time the patient finds a lump and the time he develops toxic symptoms. We now know that hyperfunctioning adenomas must reach sufficient size or volume to produce thyrotoxicosis.

Finally, there are some other patients who have been carrying innocent adenomas in their thyroids for many years, and subsequently develop diffuse hyperplasia in the extranodular tissue. This is Graves' disease. In many instances, a radioactive iodine scan will distinguish the difference between the Graves' disease with an incidental nodule and the hyperfunctioning adenomas with atrophy of the extranodular tissue. However, radioactive iodine scanning does not always show the difference in the two types of disease when there are multiple lumps present. Time does not permit us to discuss the pros and cons of using radioactive iodine to treat these lumpy thyrotoxic glands, but a great majority of these patients should be treated surgically. The objective of the surgeon in the case of the hyperfunctioning adenomas is to remove only the adenomas and preserve all the atropic extranodular tissue that is possible; otherwise, the patient will be hypothyroid for the rest of his life. In contrast, the patient with Graves' disease and an incidental adenoma deserves a very generous bilateral subtotal thyroidectomy, because the objective is to control Graves' disease. Here the removal of the masses is only for their diagnosis, while the primary objective is to reduce the excessively functioning extranodular tissue to a level that will provide a normal metabolism. In years gone by, the experienced surgeon did a generous bilateral subtotal thyroidectomy on all of his patients, because he could not tell the difference in the two diseases.

Nowadays, all patients with lumps in their thyroid that undergo surgery should have radioactive iodine in the gland at the time of the

* Picking up more radioactive iodine gram for gram than the extranodular tissue.

operation. If there is evidence of thyrotoxicosis, the presence of radioactive iodine permits the surgeon to precisely determine the disease with which he is dealing at the time the operation is in progress. It merely requires a snip of the extranodular tissue and a separate and equal snip of the tumor in a separate test tube. Both can be counted for radioactive iodine content in a few minutes while the operation is in progress. The correct operation is thus performed, because it is known precisely which tissue is producing thyrotoxicosis. Before the wound is closed, a frozen section is prepared on any cold discrete tumors that were removed just in case a carcinoma had been present.

It is of interest not only to the surgeon but also to the anesthetist to know that the hyperfunctioning adenomas are not accompanied by thyroid storm, as is sometimes encountered in a patient with poorly prepared Graves' disease. The hyperfunctioning adenomas may be removed without much, if any, preoperative preparation. On the other hand, serious calamities can be created by unsuspectedly operating on the patient with a diffuse hyperplastic gland of Graves' disease, thinking that the adenomas are to blame.

The large nodular goiter unaccompanied by thyrotoxicosis sometimes creates a problem for both the anesthetist and the surgeon. Here the issue of radioactive iodine as a therapeutic tool does not come into consideration, but trace doses are useful for identifying cold tumors and appraising substernal goiters. The large goiter is a mechanical problem which deserves surgical intervention. Actual or potential interference with the airway may pose a serious and sometimes urgent problem. The obstruction may come at one of two levels: (1) at the level of the vocal cords where paralysis of the recurrent laryngeal nerve may reduce the size of the air passage and (2) at the level of the trachea where compression or angulation by neoplasms may embarrass the lumen of the trachea. Examination of the vocal cords for evidence of paralysis of the recurrent laryngeal nerves is an important prerequisite to thyroid surgery in nodular goiter, and especially where there has been a previous operation in the region of the thyroid. The presence of a paralyzed cord in the cadaveric position suggests infiltration of the recurrent laryngeal nerve by carcinoma. Paralysis of the vocal cord in the midline position suggests that the recurrent nerve had been cut at previous thyroid surgery. The history of a change in the quality of the voice is a valuable clue to a malfunctioning vocal cord, but very often the degree of interference is scarcely detectable in patients with carcinoma, because the unaffected vocal cord may compensate by crossing the midline to effect surprisingly good apposition with the paralyzed cord in the cadaveric position. The result may be a voice which will be relatively unaffected.

The test of the patient's ability to explode a cough, however, is usually revealing. The patient can be heard to leak air as the intrathoracic pressure is raised because the chink between the cords cannot be closed tightly in preparation for the cough.

The presence of a mass on one side of the trachea may cause marked deviation of the trachea, but has little effect on the lumen of the trachea unless it is densely adherent to the trachea or there is an opposing mass on the opposite side. In the latter case, the bilateral pressure reduces the airway and produces atrophy of the tracheal rings. When the masses are removed, the trachea tends to collapse. It should be noted that most cases of postoperative tracheal collapse, attributed to soft tracheal rings, actually prove to be caused by extraluminal pressure from a hematoma. The most revealing feature of the history in a patient with compression of the trachea by a nodular goiter is the unpleasant sensation created by bending the head acutely forward, as when picking something up from the floor. Most all patients who reveal this observation or have other signs of tracheal compression have a nodular goiter which extends through the superior thoracic strait and is caught in that position.

The presence of an endotracheal tube during surgical attack on a large nodular goiter is very important. There may, however, be considerable difficulty in introducing the tube or passing it below the level of the cricoid cartilage. Experienced anesthesiologists who have handled many of these problems can usually place the tube without much difficulty, but sometimes, even then, there is some trauma produced. Some prefer to place the endotracheal tube under local anesthesia before the general anesthesia is started. If this procedure is embarked upon and does not work, a calamity may be precipitated. Some of the patients who have experienced impending airway problems refuse to be placed in a horizontal position. Often there is a large amount of thyroid tissue covering the lower larynx and trachea and completely occupying the suprasternal space so that tracheostomy cannot be done hurriedly without considerable bleeding and risk to the patient.

Where there is any likelihood that problems with the airway will be encountered and the thyroid is unusually large, I prefer to start the surgical procedure under superficial local anesthesia, raising the skin flaps, separating the muscles vertically from the hyoid bone to the sternum, and uncovering the isthmus. This is done without any discomfort to the patient while a perfectly dry surgical field is maintained. When the isthmus is divided and the cricoid cartilage and the upper most tracheal rings have been uncovered, the surgical procedure may be stopped, the wound covered, and general anesthesia started. If by chance there is any difficulty with the anesthesia, a tracheostomy in the appropriate position can be accomplished with no risk and a sterile endotracheal tube introduced so that the anesthetist may proceed with his task without intubation through the larynx. On numerous occasions when dealing with large fixed goiters, or carcinoma, I believe we have avoided considerable trouble by taking these precautionary steps first.

An appreciation of variations in the normal position of the larynx and the relationship of blood vessels to the thyroid is quite important in understanding not only the development of the substernal and intrathoracic goiter but in the surgical management of them. It is important to keep in mind that the thyroid isthmus always bears the same relationship to the cricoid cartilage in all individuals. Even though the thyroid extends deep into the chest, the isthmus will be found attached at the level of the cricoid cartilage. In the tall asthenic individual with the long slender neck, the larynx usually lies high in the neck. Several tracheal rings can usually be easily felt on physical examination, even below the level of the thyroid isthmus. In contrast, some short stocky individuals often possess a larynx which lies deep in the superior thoracic strait. Indeed, it may lie so low that the cricoid cartilage and thyroid isthmus is not palpable, because the upper edge of the sternum is in the way. If such patients develop a goiter, the fact that such enlargement is present may not be appreciated by the casual examiner. Sometimes the enlarged firm superior poles in Graves' disease may not be felt unless the examiner inserts his finger down along the side of the trachea and at the same time asks the patient to swallow. If an adenoma should develop in the inferior pole of a low-lying thyroid, it may reach sufficient size that during swallowing it cannot be raised through the superior thoracic strait. Most intrathoracic goiters develop within the chest and do not descend there, as has been postulated by some. With a dumbbell enlargement of the lower part of one lobe, and the inability of that lobe to rise on swallowing, one can recognize intrathoracic extension of a large goiter just by watching the patient swallow. The larynx does not rise freely, or there is a tendency for the larynx and goiter to rise only on one side, creating a tilting motion of the neck structures toward the side of fixation. The presence of an intrathoracic goiter is confirmed by roentgenogram of the chest and radioactive iodine scanning over this area. The radioactive iodine is of particular value in distinguishing between thyroid tissue, which always takes up some radioactive iodine, and other nonthyroidal mediastinal masses.

There has been much debate as to whether the intrathoracic goiter should be attacked through a thoracotomy incision or through the standard thyroid incision in the lower neck. The approach through the neck is the most satisfactory. The reasons are clear to those acquainted with both approaches. An appreciation of the blood vessels related to the thyroid is important. The so-called inferior thyroid arteries actually come down from a superior position in the neck, taking their origin from the thyrocervical trunk at a level almost equivalent to the superior pole of the thyroid. In many cases, there is a thyroid ima artery arising from the innominate and passing upward to reach the thyroid at the isthmus. Since this artery comes up to the isthmus, it can always be observed and carefully ligated at that level. It does not serve as a direct blood supply to the inferior poles. The first step in the attack on the substernal or intrathoracic goiter through the collar incision is to divide the superior arteries at the superior poles, roll the lobe medially, and pick up the inferior thyroid artery. This controls all of the arterial blood supply to the intrathoracic portion of the thyroid.

In considering the thoracotomy approach to an intrathoracic goiter, not only is one limited from making observations on the cervical portion of the thyroid gland, but it is impossible to attack the blood supply at its source high in the neck, or to remove much of the lesion that extends from an area in the same location. If merely the intrathoracic portion of the goiter is removed, a second operation is necessary to remove the cervical portion, which becomes adherent inferiorly as a result of the first operation.

When the blood supply has been completely controlled in the cervical approach, the veins that drain the intrathoracic portion are the only vascular connections remaining. Here, as in the case of the arteries, the veins lead upward into the neck. As one looks at the lower extension of the goiter as it passes down through the superior thoracic strait, it is the plan to grasp the edges of the thyroid capsule with clamps and gently elevate the tissue. Blunt dissection outside the capsule tends to free it from adjacent tissues and the pleura and thus make traction more effective. It is the rule never to insert the finger down along the goiter on its anterior surface, because this is the position of the veins which empty from the thyroid into the innominate veins. The goiter extends posteriorly to these great vessels. Bleeding from such veins when they are torn is very difficult to control. Thus, the intrathoracic goiter is elevated by placing the finger primarily behind the lobe and somewhat laterally to elevate it slowly into the neck. Since many of these intrathoracic goiters enlarge enormously within the chest, the large bulbous portion will not come through the narrow superior thoracic strait. With the arterial blood supply completely controlled, one can place a finger within the sac of thyroid tissue, morcellating the contents while the clamps on the thyroid capsule are used to elevate the sac. The contents are extruded and as the sac collapses, it is extracted by always keeping multiple clamps on the capsule so that traction is not lost and all thyroid tissue is removed. By this technique, several hundred grams of thyroid tissue can be removed from out of the mediastinum without bleeding, provided traction is applied primarily to the posterior part of the capsule and each vein that comes into view anteriorly is carefully divided, ligated, and permitted to drop back to its attachment on the jugular or on the innominate vein.

Just as in all other thyroid surgery, one takes meticulous care to identify the parathyroids and the recurrent laryngeal nerves. This is done before the attack on the intrathoracic portion of the goiter is begun. Because the recurrent laryngeal nerve passes around the subclavian artery on the *right,* its course is in a somewhat transverse position behind the descending thyroid lobe. It is first identified as it approaches the larynx higher in the neck and is then traced downward as the substernal portion is attacked. On the left side of the neck the recurrent laryngeal nerve comes from around the arch of the aorta, ascending parallel to the long axis of the neck and thus lying in a much more medial position to the goiter than the nerve on the right. The rule that the inferior parathyroid glands lie somewhere along the path of the lowest branch of the inferior thyroid as it passes downward to the upper tip of the thymus is an important rule to keep in mind. The thymus is very easily identified not only in children but also in elderly people during the course of any thyroidectomy. This tissue should be preserved because of the close proximity of the inferior parathyroids to it. These vessels, along with the thymus, lie in a plane to the great vessels of the neck and thus lie in a very different plane to the thyroid as it extends posteriorly and laterally into the chest.

The simplicity of removing an intrathoracic goiter through the cervical route rests entirely upon the ability to control all bleeding. The wounds are usually closed tightly, but with a catheter under vacuum for twenty-four to forty-eight hours in the mediastinum to collapse the space occupied by the goiter.

SUMMARY

There are a variety of nodular goiters, some of which deserve surgical exploration. Studies with radioautographs of lobulated and adenomatous goiters have shown which goiters should be treated by giving supplemental thyroid hormone and which should be explored. True neoplasms are not significantly affected by supplemental hormone therapy. A more objective approach to Graves' disease with incidental adenomas and hyperfunctioning adenomas, where the adenomas are responsible for thyrotoxicosis, is possible today. Airway problems associated with large goiter have been discussed and some suggestions made concerning their management. Removal of intrathoracic goiter by the cervical approach is simple and safe and most direct when the autonomical relationships are understood and the blood supply properly controlled.

Chapter 16

Insulinogenic Reserve in Unstable and Stable Diabetics and Normal Subjects

George D. Molnar, Guillermo M. Cremer, Karen E. Moxness, F. John Service, John W. Rosevear, and Eugene Ackerman

Whether, and to what extent, diabetics retain the ability to secrete insulin has both theoretical and practical importance. Theoretically, it is important to assess whether the adequacy of insulin secretory reserve (ISR) has bearing on the stability or instability of diabetes mellitus. Practically, at times it is important to determine the ISR of individual diabetic patients. This presentation will examine the problems, and their solutions, of studying ISR in insulin-treated patients. This presentation will also describe the results of a battery of tests of ISR in a series of stable and unstable diabetics and normal subjects.

Insulin has been available for forty-six years, but a sensitive, precise, and specific assay for insulin in biologic fluids has been available for only a decade. The radioimmunoassay is highly correlated with biologic activity of insulin. It is superior in many respects to the various bioassays which are also available. Much has been learned about insulin in the circulation through these assays, especially in normal subjects and in patients with mild diabetes. The increases and decreases in plasma immunoreactive insulin (IRI) concentration induced by nutrients, electrolytes, hormones, neural factors, and nonhormonal hypoglycemic substances have been intensively studied in normals and mild diabetics.

In severe diabetics, who must depend on treatment with insulin for survival, much less has been learned about their ability to secrete insulin or about the behavior of insulin in their body, whether it is insulin secreted by the patient or administered by injection. Both the native or human endogenous and the animal exogenous insulins are important, and possibly the *most* important, variables and determinants of health in diabetics. Insulin secretory ability and plasma insulin levels have not been investigated more intensively in severe diabetics because the antibodies which develop during treatment with insulin and are present in the plasma have been recognized as obstacles to the otherwise satisfactory radioimmunoassay of insulin.[1] The assay itself depends on interactions between specific antibodies and insulin.

IMMUNOASSAY OF INSULIN

The antibodies to insulin in the plasma move electrophoretically with β- and γ-globulins.[2] They associate with insulin entering the circulation and eventually dissociate from it.[2] The association creates macromolecules, confining antibody-bound insulin to the vascular compartment. The dissociation frees insulin to leave the vascular compartment, to reach tissues and act at the cellular level. The interactions of antibodies and insulin have clinical significance as well as practical utility for the measurement of both antibodies and insulin.

Insulin antibodies may be detected and measured by adding to the plasma containing the antibodies, radioactively labeled insulin and incubating the mixture.[2,3] Table 16–I illustrates the reversible reaction between

TABLE 16–I
RADIOIMMUNOASSAY OF INSULIN ANTIBODIES

		First Reaction		
AB insulin antibodies	+	RIIN labeled insulin	⇌	AB-RIIN soluble labeled complex
		Second Reaction		
AB-RIIN soluble labeled complex	+	RAHGG second antibody	⇌	AB-RIIN-RAHGG precipitated complex

RIIN = radioactive iodinated insulin.
RAHGG = rabbit anti-human-gamma-globulin.

these substances. Separation of antibody-bound labeled insulin from the free labeled insulin (not bound to antibody) can be accomplished by precipitation of the γ-globulins present. An antiserum to γ-globulin of humans can be developed in rabbits and will do this. The radioactivity in the precipitate, as a percentage of the initial total radioactivity added, indicates the proportion of labeled insulin which is bound to insulin antibodies.[3]

In normal subjects and diabetics never treated with insulin, less than 3 per cent of the initial total radioactivity is so precipitated. In insulin-treated diabetics, more than 3 per cent insulin binding by antibodies is found. Three fourths of such diabetics have 5 to 25 per cent insulin binding.[4] High antibody levels have clinical significance in being associated with resistance to the action of insulin and with chronically high daily insulin requirements. High antibody levels are also associated with delayed exit of the insulin from the circulation and hence delayed effectiveness even of intravenously infused insulin.

The radioimmunoassay of insulin evolved from the assay and studies of antibodies. The first immunologic reaction illustrated (Table 16–II) is analogous to the first reaction in the detection and measurement of insulin antibodies. Antibodies to human or pork insulin are prepared in guinea pigs. These are added, with radioactive iodinated insulin, to plasma and the expected soluble labeled complex results. The second reaction occurs when rabbit antibodies to guinea pig γ-globulin are added, to precipitate the soluble labeled complex.[5]

TABLE 16–II
RADIOIMMUNOASSAY OF INSULIN*

		First Reaction		
GPAS insulin antiserum	+	RIIN labeled insulin	⇌	GPAS-RIIN soluble labeled complex
		Competing Portion of First Reaction		
GPAS	+	IN unlabeled insulin	⇌	GPAS-IN soluble unlabeled complex
		Second Reaction		
GPAS-RIIN soluble labeled complex	+	RAGPGG second antibody	⇌	GPAS-RIIN-RAGPGG precipitated complex

RAGPGG = rabbit anti-guinea-pig-gamma-globulin.

* Usually, as in the studies reported here, the following species of insulin are used: GPAS (guinea pig antiserum) is prepared with pork insulin; RIIN (labeled insulin) is pork insulin; IN (unlabeled insulin) is human insulin.

The difference between insulin assay and insulin antibody assay is the competing portion of the first reaction. Unlabeled insulin and labeled insulin compete for binding with insulin antibodies in the first reaction. The more unlabeled insulin present (in prepared standard solutions or unknown plasma specimens), the less labeled insulin joins the soluble labeled complex. Thus, correspondingly less radioactivity is available for precipitation in the second reaction.

Figure 16–1 illustrates a typical insulin assay calibration (standard) curve. The curve indicates the decrease in antibody-bound labeled insulin as the unlabeled insulin concentration increases. Each assay has a calibration curve of this type, determined with known concentrations of insulin. Unknown plasma samples are analyzed by comparison with the known standard concentrations of insulin within the same assay.

PROBLEMS

The first problem is that of the human antibodies. As indicated in Table 16-III, these antibodies will participate in the first reaction and in the competing portion of the first reaction. Although almost always with a lower degree of energy for binding insulin than that of the guinea pig

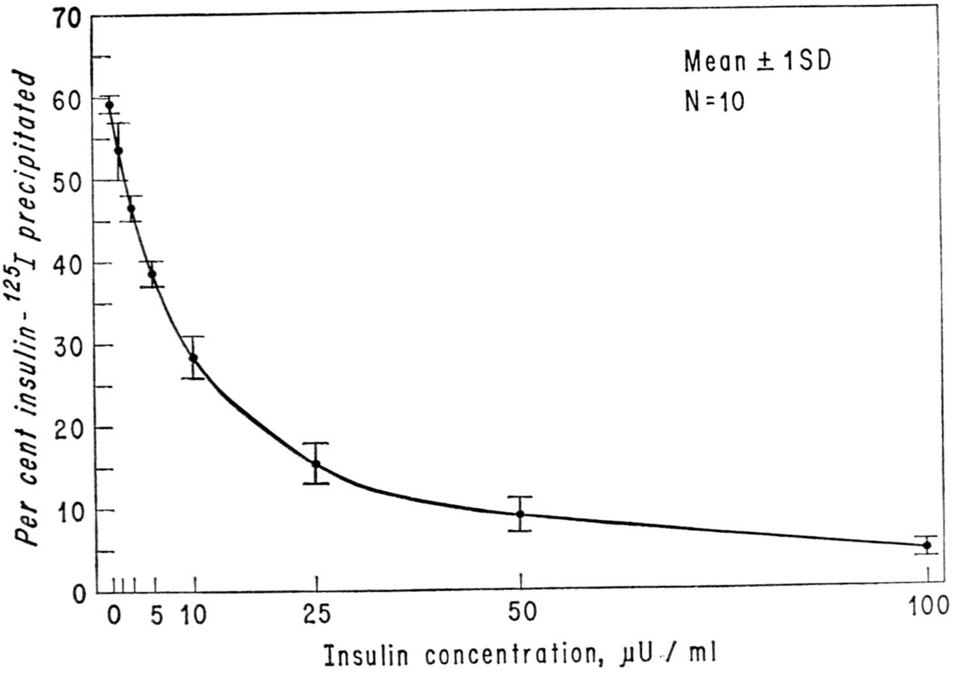

Figure 16–1. Calibration (standard) curve of immunoprecipitation assay for plasma insulin. Radioactivity in porcine insulin bound to guinea pig antibody to porcine insulin and precipitated by rabbit anti-guinea-pig-gamma-globulin is shown as percentage of total initial radioactivity in radioractive iodinated porcine insulin and is plotted against concentration of human insulin.

antibodies specially selected for assay purposes, the human antibodies may interact with both the labeled and the unlabeled insulin. Because of this competition by the human antibodies for binding insulin, the amount of insulin in the soluble complex with guinea pig antibodies may be reduced. The human antibodies do not participate in the second reaction because

TABLE 16–III
PROBLEMS OF INSULIN ASSAY WHEN ANTIBODY (AB) IS PRESENT

		First Reaction		
GPAS	+		⇌	GPAS-RIIN
		RIIN		
AB	+		⇌	AB-RIIN
		Competing Portion of First Reaction		
GPAS	+		⇌	GPAS-IN
		IN		
AB	+		⇌	AB-IN
		Second Reaction (radioactive portion only)		
GPAS-RIIN	+		⇌	GPAS-RIIN-RAGPGG
		RAGPGG		
AB-RIIN	+			No reaction

the rabbit anti-guinea-pig-gamma-globulin serum will not react with human γ-globulin to precipitate the soluble human-antibody-insulin complex.[6-8] When sufficiently energetic human antibodies are present, the percentage of labeled insulin bound to antibodies and precipitated will be decreased. As a result of this decreased percentage of labeled insulin precipitated, the apparent concentration of unlabeled insulin will be increased.

The second problem in assaying insulin in the plasma of patients who have human antibodies to insulin concerns the relative species specificity of these antibodies to the type of insulin (beef or pork) with which the patient has been treated. Such patients may have human, beef, or pork insulin present in their plasma—all three, any two, or just one. We usually know which insulin or insulins we wish to assay. We may or may not be looking for human insulin and we will usually know whether the patient had treatment with beef or pork insulin or both. It is essential that the insulin (unlabeled) used in the standard solutions to obtain the calibration (standard) curve (Fig. 16–1) be of the same species as the insulin (unknown) to be measured in the plasma specimens. In the present report, human ISR was being measured and therefore human insulin was used in the standard solutions. It helps to know also how the individual's antibodies behave toward beef or pork insulin.

Human antibodies bind beef insulin more than pork insulin, with an average ratio of 3:1.[1] Specific antibodies to beef insulin also bind pork insulin and human insulin, although in smaller amounts. Guinea pig antiserum to pork insulin will bind human insulin almost identically, but it will bind beef insulin less avidly.

In serially collected fasting plasma specimens from the same individuals, insulin binding capacity by human antibodies remains relatively constant over months and even years.[1] With increasing concentrations of added human insulin, the proportion of labeled insulin bound by antibodies which have a low insulin-binding capacity does not change, because of the low degree of energy such antibodies have for interaction with insulin. However, the proportion of labeled insulin bound by antibodies which have a high binding capacity does change because of the high degree of insulin-binding energy usually possessed by such antibodies. With increasing concentration of unlabeled insulin, more insulin is bound by energetic antibodies, although progressively smaller proportions of the total insulin available for binding is bound.

These findings have helped to explain why human antibodies, with low capacity for binding insulin, do not seem to interfere with the assay of IRI in plasma. These findings also explain how the presence of human antibodies to beef-pork insulins affects the recovery of human insulin added to plasma. Added human insulin is recovered quantitatively from normal plasma, as well as from diabetic plasma, with antibodies which have low insulin-binding capacity. However, with increasing amount of added insulin, the recovery is progressively exaggerated from diabetic plasma which has antibodies with higher insulin-binding energy and capacity.

By such *in vitro* tests of insulin binding by antibodies (and of insulin recovery), it is possible to gain better understanding of the effects of human antibodies on the assay of insulin. The validity of measurements in plasma with low antibody binding is ensured. Exaggerated apparent insulin levels may also be correctly interpreted.

To confirm these observations on the role of human antibodies to insulin, analysis of the following additional data, from diabetics with and without antibodies and from normal subjects, is in progress: (1) insulin disappearance-rate studies* with labeled and unlabeled insulin (alone and in combination), (2) insulin infusion tests during rapid and slow intravenous administration of insulin, and (3) diurnal studies with and without insulin therapy during forty-eight-hour continuous blood glucose monitoring.

Having assured ourselves of the validity of the plasma IRI measurements, my colleagues and I were ready to examine insulinogenic reserve in humans with and without antibodies.

INSULINOGENIC RESERVE

Our subjects for investigation were the same twelve diabetics and six normal individuals described in Chapter 14.[9] The design of these tests of ISR took into consideration the following train of logic. Several mechanisms of stimulating the secretion of insulin are known. Because we were examining insulin secretory ability in individuals who were expected to have impaired ISR or, in some cases, total absence of ISR, we wished to make certain, by in-depth studies of each subject, whether or not ISR was impaired and, if so, to what degree. Thus, tests with arginine, glucagon, tolbutamide, and epinephrine, as well as control experiments with saline, were done in all subjects.[10] Glucose tolerance tests were not done in severe diabetics because of the problems with the administration of large doses of glucose after withdrawal from therapy with insulin. All insulin-treated diabetics were withdrawn from long-acting insulin preparations for at least forty-eight hours prior to each test.

Since the direct measurement of insulin was complicated by the pres-

* Prolonged persistence of insulin in the circulation of diabetics[2] may lead to its accumulation and may be one of the reasons for apparently excessive IRI concentrations.

ence of antibodies in the diabetics, we wished to document the biologic effects of insulin as thoroughly as possible. Therefore, in addition to insulin measurements, simultaneous blood glucose (BG) and plasma free fatty acid (FFA) and ketone body (KB) determinations, as well as some urinary measurements, were also made.

Table 16-IV summarizes our results with the intravenous infusion of arginine (30 gm over a 30-minute period) starting after a thirty-minute base-line observation, in three different types of subjects: normal subjects, diabetics with the ability to increase their IRI concentration (these were also designated as diabetics I), and diabetics who did not increase their IRI concentration (these were designated as diabetics II). To permit comparisons, all data are calculated as percentages of the mean of the three base-line specimens (— 30 to 0 minute) which represents 100 per cent. The normal subjects had the greatest IRI increase as well as the greatest decrease in BG and FFA levels. Diabetics who responded at all had smaller insulin increases, greater BG increases, and delayed return of BG to the base line.

In Table 16-V, the effect of intravenous injection of 1 mg of glucagon is summarized, in the same manner as the data in Table 16-IV. The same individuals, in the same groups, showed the same tendencies concerning insulin secretion. Insulin response was higher in normals than in the diabetics with some ISR. A BG increase followed the increase in IRI, but only in the normals did BG return to the base line. FFA decreased similarly in normals and in those diabetics whose insulin levels also changed. Diabetics without ISR showed significant changes, compared to diabetics with ISR and to normals, only in KB and only during the first hour after the glucagon administration.

When 1 gm of tolbutamide was given intravenously (Table 16-VI), only the IRI and glucose changes were significant. The IRI increase was smaller and delayed in diabetics with ISR. BG decreased progressively in diabetics with ISR, with the lowest mean BG occurring two-and-a-fourth hours later than in normals.

When epinephrine was infused (Table 16-VII), 0.1μg/kg/min over sixty minutes, the increase in IRI followed the infusion as a rebound phenomenon. The distinction between normals and diabetics was clear. Interestingly, the relative increase in BG concentration was greatest in normals, but only the normals had regained their initial level at the conclusion of the test. The FFA patterns were similar in all three groups. However, the KB patterns differed significantly during the third hour of the test. The diabetics without ISR had the highest increases in KB levels.

TABLE 16–IV
EFFECT OF INTRAVENOUS INFUSION OF ARGININE: CHANGE FROM BASE LINE (MEAN ± SE)*

	IRI			BG			FFA			KB		
Time (min)	*Normals*	*Diabetics I*	*Diabetics II*	*Normals*	*Diabetics I*	*Diabetics II*	*Normals*	*Diabetics I*	*Diabetics II*	*Normals*	*Diabetics I*	*Diabetics II*
−30	100	103	100	102	102	101	105	95	95	104	108	97
	±2.9	±2.5	±1.2	±1.0	±1.1	±0.7	±5.7	±5.7	±7.0	±4.4	±2.3	±4.4
−15	95	99	97	100	100	100	97	99	98	96	86	99
	±3.2	±1.5	±0.7	±0.7	±0.3	±0.3	±4.7	±3.9	±6.6	±3.5	±1.9	±2.7
0	104	98	103	99	98	99	98	105	107	100	106	104
	±2.1	±1.0	±1.7	±0.4	±1.3	±0.5	±1.2	±3.1	±11.1	±4.5	±3.4	±2.1
15	425	194	101	115	105	108	75	81	105	99	101	108
	±68.3	±35.5	±4.9	±1.4	±2.4	±1.7	±6.6	±2.9	±5.7	±5.3	±6.4	±8.7
30	655	237	99	110	121	118	59	74	100	93	80	108
	±129.3	±64.8	±2.2	±4.4	±8.4	±3.1	±6.2	±6.4	±5.7	±8.6	±9.9	±8.9
45	337	212	102	97	127	126	51	77	96	84	107	115
	±66.7	±40.9	±1.6	±4.3	±8.6	±3.9	±7.0	±8.1	±5.5	±15.7	±5.2	±7.7
60	178	189	96	85	132	130	49	67	96	69	77	114
	±35.5	±40.5	±1.6	±6.0	±12.7	±4.1	±7.2	±9.8	±6.8	±16.7	±9.1	±10.9
75	157	153	97	88	129	132	52	76	101	83	71	113
	±27.4	±22.4	±1.8	±4.6	±12.8	±5.0	±7.5	±6.0	±12.0	±12.8	±8.4	±17.1
90	106	115	95	97	125	134	58	68	95	84	74	120
	±18.6	±7.9	±2.4	±5.6	±12.9	±5.3	±5.2	±8.3	±7.8	±11.5	±9.0	±16.7
105	106	118	94	98	119	133	77	69	102	89	86	102
	±18.3	±5.0	±3.2	±3.2	±13.8	±6.0	±13.8	±10.6	±9.3	±8.8	±26.3	±10.9
120	98	111	94	99	115	132	94	70.8	100	114	88	102
	±10.3	±1.6	±3.6	±4.3	±14.0	±6.6	±9.6	±13.8	±10.2	±17.1	±23.7	±16.2
135	96	102	94	103	110	132	113	79.8	97	116	106	115
	±11.6	±3.4	±4.5	±4.3	±15.9	±7.3	±20.1	±16.8	±10.3	±19.5	±33.5	±9.5
150	92	99	91	105	108	132	122	85	104	131	111	118
	±16.9	±2.2	±4.5	±4.0	±16.2	±7.7	±18.8	±14.5	±12.1	±32.2	±35.0	±9.9
165	92	93	85	107	105	133	125	83	111	162	129	151
	±12.7	±2.2	±5.2	±4.2	±17.3	±7.8	±22.0	±17.3	±10.4	±34.3	±53.2	±15.7
180	87	88	83	109	103	131	130	95	111	159	133	159
	±12.0	±4.6	±5.1	±5.0	±18.4	±7.5	±16.6	±13.1	±6.8	±35.2	±48.6	±15.0

* Shown as percentages calculated on basis of mean of three base-line values = 100%.

During infusion of 0.45% saline, with no stimulus to insulin release, IRI remained close to the base-line level in all three groups of subjects (Table 16-VIII). The diabetics with ISR had a progressive decrease in mean BG level, probably as a result of being without food. In the diabetics without ISR, BG levels were increased significantly by the third hour and their ketonemic trend was also more marked than in the other two groups.

The following summarizes our observations on ISR. By the criterion of area under the insulin response curve, normals had the greatest response in the rebound phase after epinephrine infusion. By the criterion of maximal increases of IRI, responses in normals were similar, on the average, after arginine, epinephrine, and glucagon. Diabetics with ISR had smaller proportional increases in IRI throughout. The greatest increases in diabetics occurred after tolbutamide, with arginine as a close second.

Now we are ready to combine the data on quantification of diabetic instability and on ISR in the same subjects. Table 16-IX lists the patients in order of decreasing degree of diabetic instability. The order of the patients is primarily based on the diurnal mean amplitude of glycemic excursions (MAGE), i.e. the lowest level to which we could reduce the extent of their daily average BG swings during continuous BG monitoring.[11] The order also takes into account clinical assessment. All the unstable diabetics, with one exception, had no ISR. The exception is a patient whose diabetes was discovered at age thirty-eight, fifteen years prior to these studies. She had a subnormal response to tolbutamide, a marginal one to glucagon, and none to the other tests. Among stable diabetics, the ISR was generally impaired but was clearly demonstrable.

SUMMARY

Antibodies to insulin in the plasma of insulin-treated diabetics may distort radioimmunoprecipitation assay of insulin (unless insulin-binding capacity of antibodies is low). Suitable studies of plasma containing human antibodies to beef-pork insulins *in vitro* may permit interpretation even of distorted plasma insulin concentration. Multiple tests of insulin secretory reserve were completed in five normal subjects and twelve diabetics (with plasmatic antibodies) whose degree of diabetic instability had previously been quantified. Stable diabetics had decreased but effective insulin secretory reserve, compared to normals. Unstable diabetics had virtually or absolutely no insulin secretory reserve.

TABLE 16–V

EFFECT OF INTRAVENOUS INJECTION OF 1 mg OF GLUCAGON: CHANGE FROM BASE LINE (MEAN ± SE)*

	IRI			BG			FFA			KB		
Time (min)	*Normals*	*Diabetics I*	*Diabetics II*	*Normals*	*Diabetics I*	*Diabetics II*	*Normals*	*Diabetics I*	*Diabetics II*	*Normals*	*Diabetics I*	*Diabetics II*
−10	93 ±6.6	104 ±2.1	100 ±1.0	101 ±1.2	99 ±0.4	100 ±0.3	90 ±10.8	103 ±4.1	99 ±2.8	87 ±3.7	97 ±6.7	100 ±2.4
−3	107 ±6.6	96 ±2.1	100 ±1.0	99 ±1.5	101 ±0.4	100 ±0.3	110 ±10.5	97 ±4.1	101 ±2.8	113 ±3.7	103 ±6.7	100 ±2.4
3	567 ±195.1	181 ±39.8	98 ±1.5	107 ±2.6	104 ±2.5	101 ±0.5	115 ±14.7	100 ±10.1	106 ±3.0	96 ±11.8	104 ±4.4	110 ±13.5
5	710 ±190.6	197 ±50.0	98 ±2.3	112 ±4.5	107 ±2.9	104 ±1.0						
10	261 ±56.3	180 ±44.3	97 ±2.4	131 ±4.4	118 ±4.5	106 ±1.2	123 ±27.6	101 ±11.7	115 ±3.5	116 ±11.5	93 ±0.4	132 ±6.7
20	317 ±66.8	136 ±25.0	98 ±1.4	156 ±6.7	133 ±8.4	113 ±2.4						
30	272 ±44.6	106 ±15.2	95 ±2.0	152 ±5.8	145 ±10.2	117 ±2.8	72 ±3.87	85 ±11.1	101 ±5.6	88 ±11.5	86 ±11.0	138 ±15.9
40		97 ±15.3	98 ±1.7		149 ±12.3	120 ±3.4						
50		106 ±11.9	97 ±1.9		150 ±15.3	119 ±3.9						
60	173 ±45.8	108 ±2.9	94 ±3.1	113 ±4.4	147 ±15.9	118 ±3.9		72 ±8.5	98 ±6.1		80 ±2.9	142 ±19.7
70		108 ±11.6	93 ±3.1		144 ±16.8	116 ±4.5						
80		99 ±8.5	91 ±3.2		142 ±17.2	114 ±4.2						
90	132 ±32.1	100 ±8.2	90 ±3.6	98 ±2.9	139 ±16.3	114 ±4.6	72 ±12.0	73 ±18.0	92 ±5.3	61 ±20.9	78 ±16.8	108 ±6.7
105		102 ±5.0	87 ±3.5		145 ±20.6	111 ±6.9						
120		99 ±16.5	86 ±4.7		141 ±22.6	113 ±6.9		99 ±39.6	103 ±5.0		104 ±37.6	144 ±19.7
135		104 ±1.0	85 ±3.4		138 ±24.1	112 ±15.9						
150		91 ±1.0	83 ±5.3		135 ±25.6	113 ±7.3		103 ±40.6	117 ±5.5		110 ±50.2	207 ±17.2

* Shown as percentages calculated on basis of mean of two base-line values = 100%.

TABLE 16–VI
EFFECT OF INTRAVENOUS INJECTION OF 1 gm OF TOLBUTAMIDE: CHANGE FROM BASE LINE (MEAN ± SE)*

	IRI			*BG*			*FFA*			*KB*		
Time (min)	*Normals*	*Diabetics I*	*Diabetics II*	*Normals*	*Diabetics I*	*Diabetics II*	*Normals*	*Diabetics I*	*Diabetics II*	*Normals*	*Diabetics I*	*Diabetics II*
−20	107	101	103	100	100	100	104	100	105	101	91	93
	±5.6	±1.5	±0.7	±0.5	±0.3	±0.5	±3.5	±7.3	±3.2	±8.6	±2.8	±2.4
−10	93	100	97	100	100	100	96	111	104	95	113	99
	±9.1	±1.3	±1.8	±0.4	±0	±0.2	±8.6	±10.7	±1.1	±5.7	±1.0	±2.6
−2	98	100	100	101	100	100	98	89	92	102	96	108
	±10.7	±2.3	±1.7	±0.4	±0.5	±0.6	±14.2	±3.5	±2.8	±11.6	±3.1	±3.0
10	355	142	99	98	99	101	104	86	94	74	86	96
	±82.3	±25.2	±1.8	±1.4	±0.3	±1.2	±2.4	±5.8	±9.8	±6.8	±8.5	±16.9
20	277	173	96	94	97	102	95	80	91	90		104
	±56.3	±39.9	±3.8	±2.3	±0.3	±2.0	±9.9	±6.3	±9.4	±15.2		±9.9
30	268	245	97	77	96	104	75	86	93	99	86	
	±55.4	±54.9	±3.0	±4.1	±0.3	±2.5	±5.6	±12.0	±13.0	±9.2	±12.2	
40	159	212	104	64	93	104	74	79	93			96
	±19.7	±36.8	±5.0	±5.4	±1.1	±3.2	±9.4	±7.8	±13.8			±9.0
50	159	173	98	60	90	106	84	75	91		61	
	±5.9	±23.7	±6.9	±3.6	±1.0	±3.7	±13.3	±7.7	±21.2		±14.5	
60	178	164	96	64	90	108	89	82	100	81	57	115
	±34.1	±24.1	±3.9	±3.4	±1.8	±3.9	±11.9	±4.7	±13.2	±16.7	±18.9	±11.4
90	126	162	88	74	84	111	96	87	107	97	67	134
	±9.4	±24.4	±5.6	±2.3	±2.6	±4.7	±4.8	±7.5	±18.0	±9.6	±18.7	±33.7
120	111	135	89	82	80	113	112	86	119	91	79	134
	±7.9	±10.2	±7.8	±2.7	±3.9	±5.7	±5.0	±1.9	±10.4	±8.2	±22.6	±24.4
150	95	142	90	84	76	115	114	86	112	107	77	156
	±26.0	±8.6	±6.0	±2.0	±5.7	±6.2	±7.4	±6.2	±8.1	±7.0	±16.9	±32.5
180	82	148	84	86	72	115	123	94	108	109	89	134
	±17.3	±9.2	±6.0	±2.5	±7.1	±6.3	±3.7	±3.1	±8.9	±5.9	±16.9	±31.5

* Shown as percentages calculated on basis of mean of three base-line values = 100%.

TABLE 16–VII
EFFECT OF INTRAVENOUS INFUSION OF EPINEPHRINE (0.1 μg/kg/min): CHANGE FROM BASE LINE (MEAN ± SE)*

	IRI			*BG*			*FFA*			*KB*		
Time (min)	*Normals*	*Diabetics I*	*Diabetics II*	*Normals*	*Diabetics I*	*Diabetics II*	*Normals*	*Diabetics I*	*Diabetics II*	*Normals*	*Diabetics I*	*Diabetics II*
−30	98	101	101	103	103	101	102	102	98	102	108	93
	±9.9	±0.6	±1.2	±1.8	±0.5	±1.2	±3.7	±5.9	±4.4	±6.3	±5.5	±5.6
−15	94	103	98	99	99	100	99	93	104	94	96	96
	±5.9	±2.4	±0.6	±0.5	±0	±0.4	±1.4	±2.1	±4.9	±4.4	±8.6	±5.9
0	108	97	101	98	99	99	98	106	98	103	96	112
	±14.3	±2.4	±1.1	±1.3	±0.6	±1.1	±2.4	±7.8	±5.1	±5.1	±13.9	±8.1
15	80	72	98	129	112	110	178	192	219	137	94	158
	±6.4	±0.7	±2.4	±6.1	±1.4	±3.7	±18.7	±17.2	±20.7	±19.6	±10.5	±15.7
30	76	62	100	176	137	123	275	270	303	278	168	228
	±6.7	±1.9	±3.3	±5.9	±6.0	±4.6	±29.9	±24.0	±27.7	±32.5	±21.1	±33.8
45	88	62	98	205	143	135	291	310	307	359	222	318
	±16.2	±9.0	±2.7	±5.1	±1.0	±5.4	±28.5	±28.1	±31.6	±69.1	±20.6	±47.6
60	124	67	97	226	157	147	267	298	296	415	283	376
	±19.9	±9.5	±3.1	±6.4	±3.0	±6.5	±6.7	±15.3	±33.1	±77.6	±54.7	±50.6
75	531	130	95	225	167	158	227	279	265	378	564	419
	±71.6	±35.4	±2.2	±5.1	±9.4	±10.5	±18.3	±15.3	±29.2	±78.3	±201.9	±61.5
90	611	152	92	211	170	154	149	224	192	248	386	445
	±79.4	±41.6	±2.3	±4.2	±9.0	±8.3	±17.2	±20.1	±13.0	±55.8	±55.7	±77.2
105	576	146	91	190	164	151	90	162	130	190	604	397
	±64.9	±23.0	±3.2	±5.5	±7.6	±8.6	±13.0	±11.6	±14.9	±30.9	±278.9	±84.1
120	455	135	92	165	158	150	53	122	116	133	429	362
	±36.2	±11.5	±5.1	±6.0	±6.8	±10.8	±9.1	±3.8	±13.1	±15.5	±184.7	±84.4
135	359	121	90	141	151	146	47	89	98	112	333	357
	±24.3	±18.3	±4.5	±7.0	±5.4	±11.2	±7.2	±11.3	±9.3	±14.8	±142.9	±86.8
150	311	112	87	121	145	144	43	65	83	122	169	298
	±8.9	±18.5	±4.2	±7.4	±2.1	±11.5	±8.3	±5.0	±11.6	±18.0	±60.7	±63.6
165	235	104	84	106	140	141	45	63	83	101	184	269
	±21.7	±20.3	±4.9	±6.8	±1.2	±11.9	±6.5	±2.1	±12.4	±14.0	±59.9	±59.2
180	161	110	86	95	134	140	47	66	94	101	141	249
	±19.4	±13.6	±5.7	±4.9	±2.4	±12.8	±7.1	±9.9	±19.7	±9.1	±39.9	±44.4

* Shown as percentages calculated on basis of mean of three base-line values = 100%.

TABLE 16–VIII
EFFECT OF INTRAVENOUS INFUSION OF 0.45% SALINE: CHANGE FROM BASE LINE (MEAN ± SE)*

Time (min)	IRI Normals	IRI Diabetics I	IRI Diabetics II	BG Normals	BG Diabetics I	BG Diabetics II	FFA Normals	FFA Diabetics I	FFA Diabetics II	KB Normals	KB Diabetics I	KB Diabetics II
0	98 ±5.9	99 ±1.2	100 ±0.8	101 ±0.5	101 ±0	99 ±1.0	100 ±0	100 ±0	100 ±0	100 ±0	100 ±0	100 ±0
15				98 ±1.7	99 ±0.7	99 ±0.7						
30	93 ±7.4	99 ±0.4	100 ±1.1	101 ±1.0	100 ±0.4	102 ±1.2						
45				101 ±1.6	98 ±1.0	104 ±2.7						
60	107 ±22.9	98 ±3.5	99 ±1.3	99 ±0.7	97 ±1.7	105 ±3.8	115 ±13.0	104 ±5.8	94 ±6.0	97 ±15.4	65 ±3.8	99 ±10.7
75				98 ±0.9	95 ±2.4	108 ±4.2						
90	100 ±8.5	98 ±1.8	96 ±3.9	98 ±1.8	94 ±3.1	111 ±5.0						
105				98 ±1.7	93 ±3.2	112 ±5.4						
120	91 ±9.0	98 ±2.7	95 ±3.6	100 ±2.3	92 ±4.4	114 ±5.9	112 ±13.8	110 ±7.2	99 ±5.8	97 ±15.1	92 ±7.7	111 ±8.5
135				98 ±3.3	90 ±5.4	115 ±5.4						
150	117 ±21.3	93 ±3.6	94 ±3.6	97 ±2.5	90 ±5.6	118 ±6.1						
165				100 ±2.2	88 ±6.7	120 ±6.1						
180	97 ±5.5	88 ±8.7	93 ±4.7	102 ±2.8	88 ±7.2	122 ±6.8	130 ±33.0	113 ±19.2	106 ±5.4	100 ±14.8	123 ±17.3	156 ±23.8
195				103 ±2.9	86 ±7.5	123 ±6.7						
210	85 ±7.6	89 ±9.1	91 ±5.5	103 ±2.3	85 ±8.0	124 ±6.5						
225				103 ±2.3	83 ±8.4	125 ±7.0						
240	97 ±26.2	88 ±15.3	90 ±5.5	103 ±2.1	83 ±8.4	126 ±7.4	119 ±16.5	109 ±17.8	110 ±6.4	113 ±30.0	118 ±22.5	174 ±23.6

* Shown as percentages calculated on basis of base-line value (recorded in first 30 minutes) = 100%.

TABLE 16–IX
RELATIONSHIP OF DIABETIC INSTABILITY AND INSULIN SECRETORY RESERVE

Subject	*Diabetic Instability: MAGE (mg/100 ml)**	*Clinical Appraisal†*	*Insulin Secretory Reserve‡* *Arg.*	*Gn.*	*To.*	*Ep.*
D1	198	HUS	0	0	0	0
D2	184	HUS	0	0	0	0
D3	199	HUS	0	0	—	0
D4	131	US	0	±	SN	0
D5	129	US	0	0	0	0
D6	123	US	0	0	0	0
D7	119	US	0	0	0	0
D8	119	US	0	0	0	0
D9	82	S	SN	—	LN	SN
D10	79	S	SN	SN	SN	±
D11	67	S	LN	SN	LN	—
D12	56	S	LN	SN	LN	SN

* MAGE = 48-hour mean amplitude of glycemic excursions during continuous BG monitoring. The lowest value obtained in repeated studies was used.
† HUS = highly unstable; US = unstable; S = stable.
‡ Arg. = arginine test; Gn. = glucagon test; To. = tolbutamide test; Ep. = epinephrine test. N = normal; LN = low normal; SN = subnormal; ± = marginal; 0 = none; — = not tested.

REFERENCES

1. Moxness, Karen E.; Molnar, G.D.; Taylor, W.F.; Owen, C.A., Jr.; Ackerman, E., and Rosevear, J.W.: Studies of diabetic instability. I. Immunoassay human insulin in plasma containing antibodies to pork and beef insulin. *Metabolism, 20:*1074–1082, 1971.
2. Berson, S.A.; Yalow, Rosalyn S.; Bauman, A.; Rothschild, M.A., and Newerly, Katharina: Insulin-I^{131} metabolism in human subjects: demonstration of insulin binding globulin in the circulation of insulin treated subjects. *J Clin Invest, 35:*170–190, 1956.
3. Skom, J.H., and Talmage, D.W.: Nonprecipitating insulin antibodies. *J Clin Invest, 37:*783–786, 1958.
4. Palumbo, P.J.; Molnar, G.D.; Taylor, W.F.; Moxness, Karen E., and Tauxe, W.N.: Insulin antibody binding in diabetes mellitus and factitious hypoglycemia. *Mayo Clin Proc, 44:*725–737, 1969.
5. Morgan, C.R., and Lazarow, A.: Immunoassay of insulin: two antibody system plasma insulin levels of normal, subdiabetic and diabetic rats. *Diabetes, 12:* 115–126, 1963.
6. Morgan, C.R.; Sorenson, R.L., and Lazarow, A.: Further studies of an inhibitor of the two antibody immunoassay system. *Diabetes, 13:*579–584, 1964.
7. Welborn, T.A., and Fraser, T.R.: The double-antibody immunoassay of insulin: a standardized second antibody reaction that eliminates spurious results with human serum. *Diabetologia, 1:*211–218, 1966.
8. Soeldner, J.S., and Slone, D.: Critical variables in the radioimmunoassay of serum insulin using the double antibody technic. *Diabetes, 14:*771–779, 1965.
9. Molnar, G.D.; Ackerman, E.; Rosevear, J.W.; Gatewood, L.C., and Moxness, K.E.: Continuous blood glucose analysis in ambulatory fed subjects. I. General methodology. *Mayo Clin Proc, 43:*833–851, 1968.
10. Cremer, G.M.; Molnar, G.D.; Taylor, W.F., Moxness, K.E.; Service, F.J.; Gatewood, L.C.; Ackerman, E., and Rosevear, J.W.: Studies of Diabetic Instability.

II. Tests of Insulinogenic Reserve With Infusions of Arginine, Glucagon, Epinephrine, and Saline. *Metabolism, 20:*1083–1098, 1971.

11. Service, F.J.; Molnar, G.D., Rosevear, J.W., Ackerman, E.; Gatewood, L.C., and Taylor, W.F.: Mean amplitude of glycemic excursions, a measure of diabetic instability. *Diabetes, 19:*644–655, 1970.

Chapter 17

Endocrines and Enzymes

DON H. NELSON

Before presenting a review of diagnosis and treatment of various endocrine emergencies, it must be emphasized that adrenal hypofunction following corticosteroid therapy is not associated with the signs of Addison's disease. Diagnosis of this type of adrenocortical deficit cannot be made by physical examination or by history other than by the fact that the patient has received corticosteroids in the past. Although ACTH is not useful in these patients, intravenous hydrocortisone is effective.

DIAGNOSTIC APPROACHES

Addison's Disease

One of the most important signs in diagnosis of Addison's disease is marked pigmentation and vitiligo. Base-line corticoids are usually within the low-normal range, and some sort of endocrine stimulation test is necessary. The response of urinary corticoids to the administration of ACTH is a popular test of adrenal cortex function. Although alterations in plasma corticoids concentrations can be observed following ACTH administration, at least two samples of plasma must be analyzed.

ACTH Insufficiencies

The ACTH insufficiencies of panhypopituitarism can be diagnosed by a variety of tests. The administration of pyrogen produces fever and a generalized physiological stress. The normal pituitary responds to this stress by the secretion of ACTH, and the ACTH in turn stimulates the adrenal cortex to secrete hydrocortisone. The increased secretion of hydrocortisone can be determined by measuring an increase in the urine 17-ketogenic steroids and 17-hydroxycorticosteroids. In the presence of a lesion either at the pituitary or at the adrenal cortex, the 17-ketogenic steroids and 17-hydroxycorticosteroids will not be increased after pyrogen administration. Failure of increase in these urine steroids, therefore, indicates a lesion in the adrenal cortical pituitary system but does not localize the site of lesion. In such a patient, the administration of ACTH may be helpful. If 17-ketosteroids increase following the administration of ACTH, it would imply that the adrenal cortex is intact and able to respond to ACTH. In that case, the lesion would be at the pituitary level. Another differential test is the administration of Metopirone®. Metopirone blocks the action of adrenal enzymes which are involved in the manufacture of hydrocortisone. When Metopirone has been given, the precursors of hydrocortisone (17-ketogenic steroid and 17-hydroxycorticosteroids) are formed, and appear in the urine, but the end product, hydroxycortisone, is not formed.

The hydroxycortisone is not available to circulate to the pituitary and reduce the formation of ACTH. If the pituitary is intact, it will respond by secreting ACTH and, uninhibited by the influence of hydroxycorticosterone, the urine 17-ketogenic steroids will continue to increase. If the pituitary is not intact, such ACTH stimulation will not occur and the 17-ketogenic steroids will not continue to increase.

Although it is rare, ACTH insufficiency may exist without the signs of Addison's disease, panhypopituitarism, or a history of previous steroid therapy. Patients with such an insufficiency present no physical signs or electrolyte abnormalities. Aldosterone secretion is normal. If subjected to relatively minor stress, they develop a shock-like syndrome which is fully responsive to corticosteroids.

Hyperaldosteronism

Hyperaldosteronism is a serious condition and it should always be corrected by administration of adequate quantities of potassium prior to considering induction of anesthesia. The potassium deficiency of such a patient is astounding. As much as 400 mEq K^+ deficit may occur and this cannot be corrected by infusion of 20 or 40 mEq of potassium. Replacement must proceed over a period of days before taking the patient to the operating room.

Cushing's Syndrome

Cushing's syndrome is not a medical emergency initially. The emergency arises during surgery when the adenomatous adrenal gland is removed and the remaining gland is found to be atrophied. Such an atrophic adrenal does not secrete corticosteroids during the first twenty-four hours or so following the operation. Corticosteroids can be given either intramuscularly or intravenously prior to the operation. Some prefer to give steroids intravenously during the operation and then continue with their administration thereafter. These patients require larger doses of corticosteroids than do normal individuals because their total body tissues have been perfused with large quantities of corticosteroids from their tumor over long periods of time.

These patients may undergo a difficult adjustment to normal replacement doses of corticosteroids during the postoperative period. They may develop exfoliative dermatitis and transient conditions suggestive of vasculitis or collagen disease.

A patient with the typical appearance of Cushing's syndrome has a round, red, plethoric face, abdominal striae, and multiple bruises. He may also have hypertension and diabetes. Red striae are not diagnostic of Cushing's syndrome. Red stria are seen commonly in young, obese people, particularly those who have gained considerable weight during their adolescence or early adulthood.

Hypokalemia

The most common cause of hypokalemia is diuretic therapy. Because of this, every patient who has received diuretics should have serum potassium determinations prior to anesthesia and surgery. Many hospitals perform this test automatically on admission.

Conditions other than diuretic therapy may result in hypokalemia. Patients with unilateral renal disease may develop a syndrome resembling primary hyperaldosteronism. Renin secreted by a damaged kidney may cause an increased production of angiotensin which stimulates the adrenal gland to secrete aldosterone. When the damaged kidney is removed, the aldosterone secretion returns to normal. Patients with malignant hypertension may have low serum potassium on the basis of the same renin-angiotensin mechanism. Patients with Cushing's syndrome, particularly if far advanced, may have hypokalemic alkalosis and this should be corrected prior to operation.

Finally, a large number of patients may have potassium loss due to other conditions. Vomiting and severe diarrhea may be major factors in producing potassium loss and may result in loss of large quantities of potassium which must be replaced prior to anesthesia.

Pheochromocytoma

Only one third of patients with pheochromocytoma have the typical paroxysmal hypertension, while the other two thirds develop sustained hypertension. They may have metabolic signs such as hyperglycemia or increased urinary sugar, weight loss, sweating, and diarrhea. Alterations in the urinary catecholamines, epinephrine and norepinephrine or Vanillyl Mandelic Acid may provide clues other than the clinical picture.

Many x-ray studies of the adrenal gland are done in an attempt to make the diagnosis of adrenal disease. Many of these adrenal tumors are small and elusive to x-ray visualization. Pyelograms or CO_2 inflation to outline the kidneys is useful only as a method to decide on which side the tumor is located after the diagnosis has been established by other methods.

Intravascular catheterization and determination of blood catecholamines is not generally available. If it is available, it is a useful procedure.

There are various physiological tests which are employed to diagnose pheochromocytoma. Histamine hypotension readily stimulates the secretion from the pheochromocytoma and provokes a marked hypertension. The hypertension is responsive to the intravenous administration of Regitine®. When a histamine test is done, Regitine must be immediately available to reverse the responsive hypertension, and some people prefer to do the histamine test only if there is a continuous drip of Regitine available.

Although hypertension associated with manipulation of the tumor during a surgical procedure may be catastrophic unless treated with a suitable drug such as Regitine, once the pheochromocytoma is removed, hypotension may become a major problem. Even with the administration of vasopressors of various types, patients have died in shock following the removal of a pheochromocytoma. A contributory factor in this hypotension is the chronic hypovolemia which has developed as compensation to prolonged vasoconstriction. Persistent vasospasm reduces the volume of blood necessary to fill the vascular tree. When the pheochromocytoma is removed, there then is dilatation of this constricted vascular tree with the acute fall in catecholamines and marked discrepancy in the circulating blood volume and the volume of the vascular tree. The administration of Regitine preoperatively in doses of 50 mg four to eight times daily for seven to ten days overcomes this discrepancy of circulating volume and potential vascular tree volume by dilating the vascular tree and allowing the normal reexpansion of the circulating blood volume. Even with this preparation, replacement with plasma expanders at the time the tumor is removed may be indicated.

Rarely hydrocortisone may also be needed. Most of these patients have one good functioning adrenal gland which is sufficient to sustain them following removal of the pheochromocytoma. Pheochromocytoma may be present, however, in the patient's only adrenal gland so that its removal leaves the patient without any functioning adrenal tissue, and also the patient may have received corticosteroids some time in the past unknown to present surgeons and anesthesiologists. Although normally there is no need for hydrocortisone, one must realize that in certain selected cases it will be indicated.

Panhypopituitarism

A typical case of panhypopituitarism would present the following picture: a female with a slightly low blood pressure, amenorrhea, and decrease in distribution of body hair; ketogenic steroids and hydroxycorticosteroids within normal limits although somewhat on the low side; and very low serum sodium. The low serum sodium is not due to lack of aldosterone. These patients are not sodium losers, they have severe dilutional hyponatremia secondary to their low cortisone and increase secretion of ADH which causes volume expansion. Unexplained low serum sodium may be hypopituitarism.

Hypoglycemia

There are many causes of hypoglycemia. Hopefully, patients with insulinoma are not operated on prior to their diagnosis but, nonetheless, unrelated emergency surgery may be necessary. The average patient with this condition has an insulinoma for many years prior to the time the diagnosis is made. Many of these patients are incarcerated in psychiatric institutions because the syndrome is unrecognized. The patients have bizarre mental symptoms. They may go through episodes of feeling as though they are closed off from the world. They feel that they are aware of certain things that are happening around them and that people are talking to them but they are unable to respond and are unable to act logically for the situation. These symptoms resemble those of schizophrenia or some other mental abnormality. After removal of the insulinoma, however, the patients return to normal psychiatric function.

It is interesting that although some of these patients have a normal glucose tolerance test, many of them are abnormal. They may have a diabetic glucose tolerance curve and they usually have a prolonged depression of the blood sugar in response to tolbutamide. In a normal patient, tolbutamide administration causes an acute reduction in blood sugar initially followed by recovery to normal over the next two or three hours. In a patient with insulinoma, the reduced blood sugar continues for prolonged periods. The other significant test is the twenty-four-hour fasting test. Almost all patients with insulinomas will have a precipitous and sustained fall in blood sugar when fasted twenty-four to seventy-two hours. Of course, the test is stopped when the patient becomes symptomatic and the blood sugar reaches 40 mg/100 ml. Blood sugars higher than 40 mg/100 ml, although they are suggestive, are not considered diagnostic of insulinoma.

Hypoglycemia may be due to other things. Possibly the most common cause is reactive hypoglycemia which is simply a hypoglycemia following meals. The reduced blood sugar is not maintained long enough and the blood sugar is usually not below 40 mg although rarely it may be.

Patients taking oral hypoglycemic agents may develop severe hypoglycemia just as they can after the use of insulin. Patients with either primary or secondary adrenal insufficiency may develop hypoglycemia because of an inability to carry out normal gluconeogenesis and quite often during prolonged operations they will have a fall in blood sugar. Of course, they should be given glucose during operation with either of these conditions. Finally, patients with severe hepatic disease may have a similar difficulty in maintaining blood sugar and should be given glucose.

THERAPEUTIC APPROACHES

Addison's Disease

In Addison's disease the treatment of choice, and the one which will be life-saving, is intravenous hydrocortisone.

Hypopituitarism

In the treatment of hypopituitarism, if acute, intravenous hydrocortisone is the answer.

Pheochromocytoma

In pheochromocytoma the elevations of blood pressure are likely to be due to excessive secretion of the hormone from the tumor and can usually be controlled with intravenous injection of phentolamine. Marked decreases in blood presure may occur at the time of the operation and, should volume depletion occur, replacement of volume is necessary. Some of these patients exposed to prolonged secretions of catecholamines seem to become resistant to the effect of the catecholamine. They then have persistent shock. This condition is similar to tachyphylaxis seen with other drugs. It is difficult to diagnose and is not easy to treat.

Myxedema Coma

Patients with severe myxedema may in time lapse into coma, and patients with severe myxedema who have been stressed at operation may arouse poorly. There is no good therapy for this condition. Intravenous thyroxin given in large doses may be therapeutic. Intravenous hydrocortisone is also indicated. Even with the administration of both of these substances, the mortality of the condition is high. It is preferable to make the diagnosis and correct the myxedema prior to administering anesthesia for an elective surgical operation.

Hypoparathyroidism

Hypoparathyroidism is a rare condition. Because of the very low plasma calcium level associated with this condition, patients may go into tetanic convulsions if the condition is not recognized. Diagnostically, the most useful observation is the serum phosphate and serum calcium. Almost every patient with hypoparathyroidism will have a serum phosphorus which is considerably elevated, with a markedly lowered serum calcium.

Hyperparathyroidism

Hyperparathyroidism is just the opposite to hypoparathyroidism. The laboratory findings will be those of an elevated serum calcium and a marked reduction in serum phosphorus. One of the popular currently employed therapies is the administration of phosphate solution. This will cause a marked and dramatic fall in blood calcium without the dangers which one associates with the administration of EDTA. There has been some concern that in the long-term usage of these solutions, one might have a precipitation of calcium phosphate in the tissue.

Diabetes Mellitus

The most important single factor in successfully managing the diabetic is continuous close observation. These patients can change rapidly. A juvenile diabetic may develop ketosis rapidly and requires careful examination prior to operation, during the surgical procedure, and then a careful follow-up during the postoperative period. Urinary acetone and sugar may be observed during the anesthesia administration if a catheter has been inserted, and plasma acetone may be observed also.

Chapter 18

Surgery and Anesthesia of the Thyrotoxic Patient

BROWN M. DOBYNS

When radioactive iodine became available, a new mode of therapy for hyperthyroidism began to evolve. Up until that time, bilateral subtotal thyroidectomy had been the tried and proven form of therapy for hyperthyroidism. The use of preoperative antithyroid drugs to bring the patient to a euthyroid state before surgery had reduced the surgical risk to a very low level. Before radioactive iodine therapy became popular, certain clinics and certain surgeons had become sufficiently experienced with thyroidectomy that the mortality rate in the hands of these experts was as low as one death in a thousand and the recurrence rate the world over was about 3 per cent. Although, in experienced hands, the results of surgery were excellent, the outcome in the hands of others, who did only an occasional thyroidectomy, was not always so favorable. Technical problems led to inadequate resections, which resulted in recurrence, or even persistence, of hyperthyroidism. More adequate resections were sometimes accompanied by inadvertent destruction of parathyroid tissue or damage to a recurrent laryngeal nerve. Thus, the availability of a new therapeutic tool, which was sure and was unaccompanied by the discomforts of surgery, was welcomed in many areas, particularly where poor surgery was being done. Radioactive iodine is painless. It destroys no parathyroid tissue or nerves. If it fails the first time, a second dose of radioactive iodine can be given. If it destroys the thyroid too completely, hormone supplementation is available.

For quite some time, there was fear that radioactive iodine might produce changes in the thyroid that would lead to malignant lesions, because such changes could be demonstrated in some experimental animals. However, experience has shown that the incidence of cancer of the thyroid after radioactive iodine therapy in the diffuse hyperplastic gland of Graves' disease is very low. It seems from experimental evidence that doses of radioactive iodine which are large enough to reduce the thyroid function interfere with the ability of thyroid cells to go through the process of mitosis. Thus, a failure of replication of cells

Some of the work concerning the effect of radioactive iodine is based on investigations supported in part by Research Contract AT (11–1) 1784 from the Atomic Energy Commission.

probably impedes the development of neoplasms. By the same token, experience has now shown that the failure of thyroid cells to replace themselves after radioactive iodine ultimately leads to thyroid failure many years later. Thus, although the fear of cancer from radioactive iodine has diminished, the occurrence of unrecognized hypothyroidism and myxedema has become a major problem. In such cases, there is no thyroidectomy scar to bring to mind the possibility of thyroid deficiency.

The early follow-up studies on patients treated with radioactive iodine showed that 15 to 30 per cent of the patients developed hypothyroidism within a few months. For these patients, the recognition of the complication and the correction of the deficiency was simple. Supplemental hormone was given. On the other hand, now that very long-term follow-up studies have been made ten to twenty years after radioactive iodine therapy and strenuous efforts have been made to trace all patients and not just those that were most available, it has been found that many ultimately have developed hypothyroidism. Many patients thought to be ideally treated and followed in a euthyroid state for several years, often prove to be hypothyroid. Others given thyroid supplement prophylactically have been found to have discontinued their supplement. When they find they are all right without it, they forget about the warning that was given them. When a search is made, many of those patients who were most difficult to contact and examine were incarcerated in nursing homes and homes for the aged where the nature of their disease, i.e. myxedema, was not recognized or was often regarded as senility.

Radioactive iodine is a very important tool for the treating of elderly patients or patients that are poor surgical risks. It is, however, strongly contraindicated in children or in patients who are pregnant; in fact, in the eyes of many, it is contraindicated in patients under the age of thirty-five or forty years of age, particularly in the childbearing period. Whereas radioactive iodine affects all of the thyroid tissue and ultimately causes failure, surgery preserves a small remnant of thyroid tissue, which is undamaged, and in most instances, maintains the patient in a euthyroid state. In nodular goiter where masses are present and the patient is suffering from thyrotoxicosis, surgery is indicated, because the nature of the mass or masses is not known. Surgery has a dual purpose: (1) to reduce the patient to a euthyroid state and (2) to incidentally remove the masses that are present, which in some instances may be early carcinomas. Quite recently there seems to be a trend back toward the use of thyroidectomy, for, if done well, the patient goes on through life in a normal fashion, rather than be uncertain of the outcome or be dependent on supplemental hormone the rest of the individual's life.

The administration of antithyroid drugs will control all cases of hyperthyroidism and, if given properly, should reduce the patient to a euthyroid state in five or six weeks. Unfortunately, there is a wide variation in the amount of drug required in various patients. When the patient is not improving from the administration of antithyroid drugs, there is a reason for that failure, and it should be discovered. Sometimes the patient is one who destroys the drug more rapidly than others and the dose must be increased two or even three fold to be effective. On the other hand, the patient may not be taking the prescribed dose in a regular fashion and is reluctant to admit it. In such a case, it would be a serious error for the physician to increase the dose, because the patient may then become serious about taking the medication regularly and would receive a dose that is toxic for him. When the patient is taking the prescribed dose, it is very easy to give a trace dose of radioactive iodine and see if the original high uptake of radioactive iodine is being blocked by the amount of drug that the patient is taking, or claims to be taking. The dose is then adjusted to a level that is effective, and within four to six weeks all patients should be under control.

The administration of an antithyroid drug is not always necessary as a preoperative measure in patients with Graves' disease. Before antithyroid drugs became available, iodine administration was a standard form of preoperative medication. Since many patients may be operated on in a perfectly safe fashion after taking only iodine for a few days, this form of preoperative preparation is quite satisfactory in mild thyrotoxicosis and requires much less preoperative time. Unfortunately, it is sometimes difficult to judge just which patient will require antithyroid drugs and which will require only iodine. If iodine has recently been given, any further benefit from it is unlikely. There is often a delay in improvement from antithyroid drugs if the gland is loaded with iodine first. Thus, if there is any doubt in the physician's mind, then it is appropriate to start the antithyroid drug first. After the patient has established the habit of taking the drug regularly over several days, saturated solution of potassium iodide may be introduced. The iodinating effect on the thyroid occurs, even though the uptake of the iodine by the gland is blocked.

Thus, if the iodine is given in addition to the antithyroid drug and a dramatic benefit is derived from the iodine promptly, there is no reason to wait for the effect of the drug. The operation can be performed promptly. If the patient does not derive benefit from the iodine, the antithyroid drug should have been blocking the gland and the improvement from the antithyroid drug should have been taking place so that there would have been no delay. If the iodine therapy had been tried alone first and failed, the gland would have been loaded with iodine and there would have been a considerable lag time in the improvement when antithyroid drug was given.

Occasionally, surgeons have gotten into trouble by operating on a patient in whom they had stopped the antithyroid drug and had given iodine before surgery. In the belief that it was important to reduce the vascularity of the gland by giving iodine, some have reasoned that the involuting effect of iodine could not take place unless the iodine was taken up by the gland. In patients who had been severely toxic, the discontinuation of the drug for ten days to two weeks allows for some recrudescence of the disease and the surgery is then carried out at a time when the patient is becoming toxic again. Since the iodinating effect takes place even though the thyroid uptake is blocked, one does not stop the antithyroid drug if one feels the need to give iodine immediately before surgery.

One of the most serious errors in the preparation for surgery with antithyroid drugs has been overtreatment of the patient. This is particularly true if there is a prolonged period of preoperative preparation. If the patient's metabolism comes down very slowly, or if the patient is not watched closely, he may be completely euthyroid before plans for surgery are made. With the delays in admissions to hospitals these days, the antithyroid therapy may be continued for an extra two or three weeks. When the patient is admitted, he may actually be in a hypothyroid state. Sometimes a delay is created by certain laboratory testing to "make sure" the patient is ready for surgery. Actually, the clinical judgment of the patient's status is the best basis for appraising the time for surgery. This judgment is based on observations such as (1) disappearance in the lability of the pulse with a decline in the resting rate to 85 or below, (2) loss of the tremor, (3) a recovery of muscle strength, and (4) a gain in body weight.

Hypothyroidism at the time of surgery may be a serious error. In practice, things seem to go best if the patient is still very mildly toxic. Patients who are hypothyroid are overly responsive to sedation or preoperative medication. They take the anesthesia in a notoriously poor fashion. Anesthesia in patients for thyroidectomy should be very light. The patient should be essentially awake at the time that the procedure is completed. The hypothyroid patient is sluggish and responds poorly during the recovery phase. He does not actively clear his respiratory tree, and he is very much disinclined to respond to the encouragement to get out of bed and walk around the evening of the operation. He may be inclined to stay in bed for a number of days following surgery. All of these features are serious handicaps to getting the patient out of the hospital in three or four days and back to work in a week or ten days.

The preoperative preparation for toxic adenoma is somewhat different from that for Graves' disease. Thyroid crisis or storm is a physiological phenomenon associated with Graves' disease and not associated with toxic adenoma. Where adenomas are present and are to blame for the thyrotoxicosis, there is little need for preoperative antithyroid drug therapy unless the patient is in a debilitated state or suffering from cardiac disease. It should be acknowledged that if an antithyroid drug is needed in these patients, a longer period of treatment is usually required in this disease than in Graves' disease.

Exophthalmos is a problem in some patients with Graves' disease. Quite some years ago, the idea was proposed that correction of the hyperthyroidism resulted in an excessive output of TSH from the anterior pituitary. At that time, it was thought that TSH was the cause of exophthalmos. Some features of the disease led to this conclusion. Occasionally, the exophthalmos of Graves' disease continues to progress, even after the hyperthyroidism is corrected. Furthermore, the deficiency in circulating thyroid hormone associated with hypothyroidism results in a waterlogging effect that is superimposed on the edema and swelling about the eyes. This added edema makes the exophthalmos worse. Just as the patient with incompetent saphaneous veins in his legs develops edema after standing all day with his legs in a dependent position, so the recumbent posture for a patient with Graves' disease increases the edema about the eyes when he lies down. This is why patients with progressive exophthalmos have much more trouble in the morning than they do in the afternoon and evening after they have been upright all day. Thus, a patient put down in a horizontal position during surgery and during the postoperative recovery period may display an increase in edema about the eyes. Special attention should be given to protecting the cornea, and the tight face mask should be avoided during anesthesia. The patient should have the head elevated in the postoperative period. Rather than avoid correction of hyperthyroidism, as advocated by some, it now seems clear that the eyes most often improve when hyperthyroidism is corrected. There is no difference in the subsequent course of exophthalmos after thyroidectomy or radioactive iodine therapy.

The principles of heavy sedation and large doses of narcotics as preoperative medication for the patient with thyrotoxicosis is no longer important, because patients are not operated on until they are near the euthyroid state. Physiologically, these patients are normal. Small amounts of sedation are good. Considerable caution should be exercised to avoid general anesthesia in patients who have been overtreated with antithyroid drugs, and especially after being given heavy sedation, particularly morphine and neuroleptic drugs. Whereas it was customary in years gone by to operate on many of these patients with local anesthesia so that they might be talking to the surgeon and thus guarantee the integrity of the recurrent laryngeal nerves during the course of the procedure, this has given way to preference for general anesthesia and meticulous care to see and avoid the recurrent laryngeal nerves during the surgical procedure.

I should like to emphasize meticulous hemostasis in thyroid surgery. There is no substitute for tying every small vessel that leads to or from the thyroid gland as the dissection proceeds. Any clamp that comes off of a vessel, even though the vessel does not bleed, should be replaced and a tie placed, because ultimate pressure in that vessel may lead to bleeding. No thyroidectomy should result in more than 200 or 300 cc of blood loss. If there is more loss than this, it is an indication that the surgeon is trying to go too fast. It is a reasonably good possibility that he will damage the parathyroids if he has much bleeding. All parathyroids should be clearly identified and preserved on their vascular pedicles. The recurrent laryngeal nerves should be seen on both sides of the neck and the lobe mobilized in such a way that the entire course of the nerve is carefully noted. It seems appropriate to divide the isthmus first so that the lobes might be more easily mobilized and slid forward on the trachea in such a fashion as to see clearly the entire posterior surface of the lobe and its relationship to the other vital structures. In all cases of Graves' disease, the thymus is enlarged, and in the surgical procedure for Graves' disease, one most always sees the upper tip of the thymus lying posterior to the inferior pole of the thyroid. The inferior branch of the inferior thyroid passes downward to supply the thymus. The inferior parathyroids are almost always identified along this vessel. It is extremely important to ligate the superior pole vessels above the thyroid tissue. Any tissue allowed to remain on the superior pole vessels may hypertrophy and make a conspicuous mass in the neck. Similarly, it is important to remove the pyramidal lobe which leads from the isthmus up to the hyoid bone. It is hypertrophied in most patients with Graves' disease. Subsequent further enlargement of this structure presents as a marble-sized mass perched on the larynx. The small remnant of thyroid lobe left next to the trachea at a point where the nerve enters the larynx has all vessels on its bleeding surface carefully tied, but the continuity of the inferior thyroid artery leading to this remnant is carefully preserved. Placing a ligature around this artery to control bleeding is a very insecure way of managing a bleeding problem, because bleeding through collateral vessels often takes place even though the main flow is interrupted. A more important consideration is the increased incidence of hypothyroidism in the postoperative period when this main blood supply to the remnant has been interrupted.

Surgeons that are acquainted with the pathology of Graves' disease are keenly alert to the gray crumbly character of some of these thyroids. This is a manifestation of thyroiditis, or perhaps we should say lymphocytic infiltration. When a surgeon encounters such a gland, it is important to allow much more tissue to remain, because one can predict that if a standard resection is done in a patient with considerable lymphocytic infiltration, the incidence of hypothyroidism, indeed myxedema, is very high. This postoperative complication can be greatly reduced by this precaution.

It is a foregone conclusion that all patients should have a vocal cord examination by indirect laryngoscopy prior to anesthesia. Thyroidectomy cannot be done safely without observing both recurrent laryngeal nerves. Handling the nerves with instruments will cause temporary paralysis. Trauma is avoided by handling only the tissue near the nerve as the dissection progresses. If a postoperative check of the vocal cords is done, it should be done within an hour or so after surgery when the patient is awake but before there is edema developing in the operative site. Some swelling of tissue near the nerve may mildly impair the function of it and give a misleading impression of paralysis.

Postoperative complications can be reduced to a very insignificant level if precautions are taken. Postoperative bleeding is avoided first by placing ties on all blood vessels regardless of their size and regardless of whether or not they bleed. Such vessels may bleed later. It has been my policy to induce the patient to strain by moving the endotracheal tube at the conclusion of the resection and before the wound is closed. A good vigorous Valsalva maneuver will cause veins to open and bleed. These can be ligated at the time, rather than have them bleed when the patient coughs and strains after the wound has been closed. Having tested the patient in this fashion, I prefer to close the wound tightly in layers. Placing a bouquet of drains in a thyroidectomy wound is no substitute for careful ligation of all vessels. Very often, if bleeding takes place in a wound in which there are drains, blood clots form in such a fashion as to obstruct the egress of the blood. Thus, pressure is built up, even though drains are in place.

Postoperative headaches following thyroidectomy are common. There are at least two causes of this. Both can be avoided. The neck may have been unduly extended during the operation. The patient often has a tendency to hold his head in a rigid fashion after operation because of the discomfort in the operative area. The tight muscles of the neck ache and pull after twenty-four hours. He should be reminded to exercise his neck freely from one side to the other. In this way muscle spasm is avoided.

A word should be said about the treatment of thyrotoxicosis in pregnancy. The importance of avoiding radioactive iodine therapy here is a foregone conclusion. Some physicians have preferred to treat patients through the entire period of pregnancy with antithyroid drugs. This can be done if therapy is very closely supervised. However, injudicious use of antithyroid drugs often leads to a goiter in the infant, which, on rare occasions, may have some pressure effect on the trachea. Probably the best method of therapy of thyrotoxicosis during the first trimester of pregnancy is temporary treatment with antithyroid drugs. Bilateral subtotal thyroidectomy may be done during the second trimester and is tolerated well and accompanied by no increase in the rate of abortion. If the patient develops thyrotoxicosis late in pregnancy, control of the disease with antithyroid drugs until the pregnancy is terminated seems wise. Subtotal thyroidectomy can be performed a few weeks after delivery. Anytime during pregnancy when the hyperthyroidism is controlled but surgery is delayed until later, the dose of antithyroid drug should be reduced so that the patient remains mildly toxic until definitive therapy can be undertaken.

Infants born of thyrotoxic mothers are usually thyrotoxic for the first few days of life. They very frequently show exophthalmos, sometimes to a striking degree. There is no therapy necessary in these infants, for the disease very rapidly disappears.

Graves' disease in children has been thought by some to pose more of a problem of management than it does in adults. Children with mild hyperthyroidism can sometimes be controlled with iodine therapy alone. Most frequently this is only of temporary benefit, and as the disease progresses the patient begins to show increasing evidence of toxicity in spite of the iodine. If the disease is not controlled readily with iodine, the patient should be prepared for surgery and a bilateral subtotal thyroidectomy done. The patient should miss no more than ten days of school from the time the thyroidectomy is performed.

Occasionally a patient with an acute surgical emergency is discovered to be suffering from a previously unrecognized and untreated Graves' disease. Quite naturally there is a great fear that a surgical procedure will precipitate a thyroid storm. In actual practice there are several factors which make the gravity of the situation less dismal than it might seem. First, the more severe the thyrotoxicosis, the less likely we are to encounter a serious problem, because the disease would have been recognized and at least some therapy would have previously been given. If any therapy had been given, even though the patient is still thyrotoxic, there is much less likelihood of precipitating a storm. Second, a surgical procedure, if not too traumatic, may reduce the aggravation of the thyrotoxicosis. Third, there is usually more opportunity to procrastinate about doing the surgery for hours or even days than might seem appropriate when first confronted with this complex problem. Procrastination gives time to get some protective therapy underway, but unfortunately, there is no drug that brings thyrotoxicosis under control immediately. Radioactive iodine requires weeks. Antithyroid drugs require days; and a steroid, although it will usually modify thyroid storm, may be premature or contraindicated in the face of the other disease. Iodide gives some benefit in hours and certainly in twenty-four hours. The pulse slows, the bruit disappears, the restlessness is less, and a storm may be thwarted. If possible, one or more large doses of antithyroid drugs can be given shortly preceding the iodide to raise the blood level of the blocking agent rapidly in case it can be continued. It is surprising how procrastination over the eminence of surgery pays off, unless there is clear evidence of a true emergency, such as hemorrhage or rapidly advancing severe infection, in which case, simple drainage of an abscess may make the patient better able to withstand the thyrotoxicosis. If surgery is indicated, it should be confined to the simplest possible procedure. The principles that were learned before antithyroid drugs became available when thyroidectomies were performed on thyrotoxic patients, still hold true, i.e. heavy sedation, local anesthesia, and avoidance of general anesthesia if possible. The prophylactic use of antibiotics is of great value in protecting against pneumonia.

A somewhat similar complex situation arises when myxedema is discovered at the time of an impending surgical procedure. Here, fortunately, we have triiodothyronine, a rapidly acting form of thyroid hormone, which will raise the patient's metabolism rather promptly. Here also, cautious procrastination concerning surgery is useful. More often than not, myxedema is unfortunately so subtle that it is sometimes not recognized in the presence of the other disease and a respiratory calamity results from the liberal use of preoperative medication before the surgical procedure gets underway.

In conclusion, there is a place for radioactive iodine therapy for Graves' disease in older patients or poor surgical risks. The fear of carcinoma of the thyroid following radioactive iodine therapy is no longer a serious threat. The presence of late-developing unrecognized hypothyroidism and myxedema is a serious problem following radioactive iodine. It is frequently unrecognized in elderly patients and in those individuals whose intelligence is relatively low. Surgery is the preferred form of therapy in young individuals, in women during pregnancy, and in adults in the chlid-bearing period. It is also preferred in thyrotoxicosis where nodules are present in the thyroid. Surgery should be preferred where it is important to have a remnant of thyroid tissue that will maintain the patient in a euthyroid state the rest of his life. It is unfortunate that there has been some decline in the training of residents in the surgical management of thyroid disease.

Chapter 19

Transphenoidal Surgery

THEODORE ROBERTS

Until the early 1920's, the transphenoidal route to the sella turcica was the method of choice for the surgical approach to the pituitary gland. A gradual transition of surgical methodology then occurred so that most neurosurgeons preferred the frontal craniotomy and a subchiasmatic approach to the pituitary gland. Our gifted predecessor in neurosurgery, Doctor Harvey Cushing, had carried out transphenoidal surgery for resection of pituitary tumors in 74 per cent of 338 cases of pituitary adenomas: he reported remission of disease in some patients lasting up to thirty years.[1] Doctor H. Hamlin[2] reported a mortality of less than 2 per cent in 104 patients operated on for pituitary tumor excision employing the transphenoidal surgical approach. Doctor Jules Hardy, Notre Dame Hospital in Montreal,[3] in a group of twenty consecutive patients treated by the transphenoidal technique reported no operative mortalities.

What might be described as "parasurgical advances" have led us again to the transphenoidal method of pituitary gland excision. As is often the case, advances in other related and, in some instances, nonrelated fields have been obvious precursors to progress in medicine. In particular, I would refer to the development of a portable radio fluoroscopic control method employing an image intensifier and a television monitoring system in the operating room. This equipment allows the surgeon to directly observe the placement of drills and probes in the region of the anterior wall of the sphenoid sinus and the floor of the sella turcica, an otherwise relatively blind approach when one is employing the transphenoidal surgical procedure.

Another adjunctive measure which is of considerable importance for the adequate performance of this type of surgery has been the availability of an operating room microscope with a high-intensity built-in illumination system that can be directed through a small port into the depths of a surgical wound. Other developments worth comment are the miniaturized and motorized hand drills and hand instruments allowing the surgeon more precision in the handling of remote tissues seen in microscopic fields. One should refer to the excellent endotracheal anesthesia methods now safely employed in the debilitated patient that is not infrequently the candidate for a procedure on the pituitary gland such as a hypophysectomy. Another adjunctive measure which relates to the success of pituitary gland procedures is that of available endocrine replacement therapy and in particular the now readily available cortisone and its immediate derivatives.

Guiot Thidaut[4] in 1959 introduced the method for removal of the pituitary adenoma via the transphenoidal method under radiofluoroscopic control. Jules Hardy[5] then further defined this technique aided by the stereoscopic microscope and has, since 1962, recommended the use of this technique for resection of intrasellar lesions and, in some instances, where the lesion extends above the level of the diaphragma sella, such as the craniopharyngioma. He has used this procedure in total hypophysectomy as a means of palliation of the patient with metastatic carcinoma of the breast and in carefully selected patients with diabetic retinopathy. His operative approach ("oro-nasal midline rhinoseptal transphenoidal") has the specific advantage in that it allows the direct visualization in minute detail of the operative site through the dissecting microscope. X-ray observation, obtained in the operating room with the use of an image intensifier and an adjacent television monitor, allows for direct observation of the exact position of the bone drills, chisels, and dissecting probes. Hardy pointed out that with this technique a reduced incidence of rhinorrhea exists if the surgeon places an annular-shaped piece of septal cartilage into the opening of the sphenoidal sinus after packing the sinus with muscle tissue.

Disease states where the method of transphenoidal hypophysectomy may be used are generally in three categories: sellar tumors (adenoma, craniopharyngioma, chondrosarcoma); palliation for metastatic carcinoma (breast and possibly prostate carcinoma); and in hormonally influenced disease states such as diabetic retinopathy, acromegaly, and Cushing's disease.

TUMOR EXCISION

The transphenoidal route is employed in a number of methods such as a stereotatic approach with cryo probes or less commonly employed[6] yttrium-90 seeds that are placed within the lesion and also direct metallic probes which can then be used for tissue coagulation. It would appear that there are advantages to direct visual approach to the pituitary gland when the resection is being carried out under the operator's close scrutiny. This technique then can perhaps allow for a lesser risk of damage occurring to the surrounding structures such as the optic nerves or internal carotid arteries. Lesions with suprasellar extension may be approached following prior pneumoencephalography or ventriculography to fill the third ventricle and allow the operator to directly observe changes in ventricular configuration as the tumor mass is withdrawn downward from its suprasellar position. If the lesion is principally suprasellar in extent, then a transfrontal subchiasmatic approach would be more satisfactory. With the direct transphenoidal surgical approach and in particular with the intrasellar lesions one might anticipate a more complete tumor resection with the minimal hazard of optic nerve damage. The optic chiasm position relative to the sella turcica often presents a challenge to the surgeon when it is in the so-called prefixed or postfixed position. These anatomical variations of optic nerves pituitary stalk relationship tend to further decrease the surgeon's access to the sella turcica.

PALLIATIVE HYPOPHYSECTOMY

Since the observations of Beatson in 1896, that alterations in the endocrine state of the body have salutary effects for control of some carcinomata, numerous studies have been attempted to learn the true efficacy of "hormonal deprivation treatment" with tumors thought to be dependent on hormonal support. After 1950, when cortisone became available, adrenalectomy and subsequently hypophysectomy procedures were tried. Doctor Bronson Ray[7] has commented on the mammotrophic hormone, prolactin, as likely to play a dominate role in breast carcinoma growth. Further he stated that estrogen is important only in its role as a physiological stimulant of the mammotrophic cells of the anterior pituitary gland. Growth hormone is suspect as a tumor stimulant, and unfortunately the results of hypophysectomy for tumor involution appear rather inconstant and the period of palliation remains indefinite. Mortality figures for hypophysectomy via transfrontal surgical approach for carcinoma of the breast have been listed at 7 per cent in earlier series and now are reported as low as 1 per cent with the transphenoidal route. Doctor Ray reported[7] that 42 per cent of the patients with hypophysectomy and carcinoma of the breast had remissions of six months or more. In an attempt to establish an indicator of which patients with metastatic breast carcinoma will improve with the hypophysectomy, Atkins, *et al.*[8] in 1960 initiated studies of the androgen and the corticosteroid metabolites found in urine and established what they have described as a "discriminate function factor." Those patients with "positive discriminate functions" were recommended for hypophysectomy directly and those with "negative discriminates" were considered better candidates for adrenalectomy. In this ten-year study it was noted that with an increased urinary excretion rate of androsterone and etiocholanolone, there was a better response of these patients to pituitary gland destruction. They found that the patients with a previous mastectomy responded about as well to adrenalectomy or hypophysectomy, whereas those without mastectomy improved more reliably with the hypophysectomy.

Diabetes mellitus patients are on occasion considered for hypophysectomy when they have evidence of progressive diabetic retinopathy constituting a threat to the patient's preservation of vision. Why some 78 per cent of these patients, as has been described by Ray,[9] will have visual improvement posthypophysectomies is as yet not clear. Vasculopathy of the retina is seemingly not related to that of other vascular structures as is seen in the kidney of the diabetic patient. A devastating feature of the retinal disease is that of blood oozing into the vitreous from microaneurysm sites and the occurrence of retinal vein occlusions.

Diabetic retinopathy patients being considered for hypophysectomy must be carefully selected and require thorough preoperative screening by an endocrinologist. The mortality rate reported by Luft[10] and Olivecrona in 1955 was 33 per cent, and notable reduction in this mortality rate to around 1 per cent is now reported by J. Hardy. The improvement apparently is due to the methods of patient selection and in part the decreased morbidity of transphenoidal approach. Patient-selection criteria have been mentioned by Ray[9] and others and are as follows: The patient should be forty years or younger; should have useful vision in one eye; evidence of progressive visual impairment; limited degree of retinitis proliferans; adequate renal function (a BUN lower than 30 mg/100 ml), creatinine clearance of greater than 50 cc/min; urea clearance of greater than 50 per cent of normal, and albumin urea of less than 5 gm/24 hr. There should be no evidence of chronic pyelonephritis and the patient should have adequate cardiovascular function. Hypertension may be present but with a diastolic pressure of less than 100 mm Hg. They would not be candidates for hypophysectomy if there is a history of angina, myocardial infarction, cardiac decompensation, or severe peripheral vascular disease. These patients should be reasonably intelligent individuals capable of following postoperative instructions regarding insulin therapy, etc. Ray reported that in his last twenty-one transfrontal hypophysectomy cases which were selected with the above criteria that no postoperative deaths occurred. Following hypophysectomy, retinal neovascular findings such as microaneurysms and macroaneurysms, pre-retinal and retinal hemorrhages showed evidences of regression. Retinal exudates may be observed to gradually disappear. Cogan,[11] lending a cautious note, suggested that pituitary resections be reserved for the few cases where reasonable certainty of blindness will occur if the procedure is not carried out. A juvenile diabetic with proliferative retinopathy carried a 50 per cent risk of blindness occurring within two to three years. Other criteria for patient selection as candidates for hypophysectomy have been recently mentioned. Kapur *et al.*,[12] performing transphenoidal hypophysectomy for metastatic carcinoma of the breast, stated that any patient who is considered fit for general anesthetic would be acceptable for this procedure.

This review of the transphenoidal methodology of intrasellar removal of the pituitary gland or tumors suggests that these individuals, often quite debilitated, can, with newer parasurgical methods available, undergo this type of surgery with a low mortality and morbidity.

REFERENCES

1. Henderson, W.R.: The pituitary adenomata. A follow-up study of the surgical results in 338 cases. *Br J Surg, 26:*811–921, 1939.
2. Hamlin, H.: The case for trans-sphenoidal approach to hypophysial tumors. *J Neurosurg, 19:*1000–1003, 1962.
3. Hardy, J.: Trans-sphenoidal surgery of pituitary fossa tumors with televised radiofluoroscopic control. *J Neurosurg, 6:*612–619, 1965.
4. Guiot, G., and Thibaut, B.: L'extirpation des adenomes hypophysaires par voie trans-sphenoidale. *Neurochirurgie, 1:*133–150, 1959.
5. Hardy, J.: La chirurgie de l'hypophyse par voie trans-sphenoidale; etude comparative de deux modalite techniques. *Union Med Can, 96:*702–712, 1967.
6. Gelinas, J., *et al.*: *Cleve Clin Q, 33:*171–175, 1966.
7. Ray, B.: Hypophysectomy as palliative treatment. *JAMA, 200:*166–167, 1967.
8. Atkins, H., *et al.*: Urinary steroids in the production of response to adrenalectomy or hypophysectomy. *Lancet, 2:*1261–1263, 1968.
9. Ray, B., *et al.*: Pituitary ablation for diabetic retinopathy. *JAMA, 203:*79–84, 1968.
10. Luft, R., *et al.*: Hypophysectomy in man. Further experiences in severe diabetes mellitus. *Br Med J, 2:*752–756, 1955.
11. Cogan, D.: Diabetic Retinopathy. *N. Engl J Med, 270:*787–788, April 9, 1964.
12. Kapur, T., *et al.*: Trans-sphenoidal hypophysectomy for metastatic carcinoma of the breast. *Br J Surg, 56:*332–337, 1969.

Chapter 20

Exaggerated Hormonal Responses to Anesthesia

A.H. Giesecke, Jr., and M.T. Jenkins

The endocrine-hormonal response to the anesthetic experience begins in the preoperative period. The most potent stimuli to sympathoadrenal secretion of epinephrine and norepinephrine are psychologic stimuli such as excitement, fear, or pain. Anoxia, hemorrhage, hypovolemia, respiratory acidosis, and some types of anesthesia also represent potent stimuli to sympathoadrenal hyperactivity.

The anxious patient with sweaty palms, blanched skin, tachycardia, and systolic hypertension will require a careful preoperative interview to discover the source of his anxiety, a reassuring description of his expected operative and postoperative course. The preoperative interview can in this way represent a major hormonal event in the patient's hospital stay. Egbert and Beecher[1] in a blind study showed that patients who had received a preoperative interview by their anesthesiologist were less nervous than patients who had received placebo or pentobarbital premedication without an interview.

Martinez and Norlander[2] found that premedication with morphine, morphine-hyoscine, pentobarbital-atropine, hyoscine, or placebo was not effective in suppressing the preoperative "stress" phenomenon measured by urinary excretion of epinephrine and norepinephrine. However, they found only four of thirty patients with E/Ne excretions above 20μg/min, which is the rate of excretion considered to be indicative of significant emotional tension.

The sleep produced by anesthetic agents is not physiological sleep, nor is it restful. General anesthesia is a stress situation which sets off an "alarm reaction." In man, the effect of anesthesia on plasma catecholamine concentrations has been extensively studied. Anesthesia with thiopental-nitrous oxide or halothane causes little or no change in plasma catecholamine levels. Cycloproprane and diethyl ether produce a marked increase which is predominantly norepinephrine. Epinephrine is less consistently elevated. It is therefore concluded that the sympathetic stimulation involves mostly extra-adrenal sites which tend to release more norepinephrine than epinephrine.

The rise of plasma norepinephrine is a direct correlation to the depth of anesthesia with cyclopropane. The higher the inspired concentration of cyclopropane, the higher the plasma level of norepinephrine. This correlation does not exist with other anesthetic agents. The norepinephrine level during ether anesthesia is variable over a wide range of ether concentrations. The sympathetic overactivity can be blocked by high spinal anesthesia.

Anoxia and respiratory acidosis stimulate the secretion of E/NE. The sympathetic response to anoxia and respiratory acidosis is retained during general anesthesia but blocked by spinal anesthesia. Most of the metabolic effects of respiratory acidosis can be explained on the basis of sympatho-adrenal stimulation.

Innovar-nitrous oxide anesthesia is associated with an elevation of the urinary excretion of *epinephrine* rather than norepinephrine. In a study of twelve patients,[3] the excretion of epinephrine in urine per hour was found to be significantly elevated during a period of anesthesia without surgical stimulation. This finding contrasts with the lack of change during cyclopropane or halothane anesthesia. The adrenal medullary stimulation under these circumstances may explain the cardiovascular stability seen during neuroleptic analgesia, including the occasional episode of hypertension seen.

Stimulation of the sympathetic nervous system is important in the homeostatic adjustment of the circulation to anesthesia. All anesthetic agents when given in isolated-denervated circumstances tend to produce myocardial depression and vasodilatation. This would indicate that sympathetic stimulation is partially responsible for the cardiovascular stability seen with ether and cyclopropane. Conversely, the absence of stimulation of the sympathetics may allow the development of hypotension and bradycardia seen in halothane, methoxyflurane, and spinal anesthesia.

The mechanism of stimulation of the sympathetic nervous system is related to an increased activity in the medullary vasomotor center plus, in the case of cyclopropane, a sensitization of the receptor site to endogenous E/NE. On the other hand, the failure of halothane to stimulate the sympathetic nervous system seems to be related to the ability of halothane to depress synaptic transmission in the ganglia. Miller and Briscoe[4] have demonstrated a 25 per cent increase in impulse traffic in preganglionic cervical splanchnic and adrenal sympathetic nerves with cyclopropane, ether, *and halothane* anesthesia.

Jenkins, Livingston, and Crout[5] have shown in isolated bovine adrenal glands that halothane will depress the elaboration of catecholamines in reponse to stimulation by acetylcholine in the bath. The effect is one of inhibition of the stimulus-secretion coupling and is possibly related to inactivation, or inhibition of ionized calcium. This is probably the same mechanism by which anesthetics depress muscular activity, myocardial contractility, and other excitation-contraction or stimulus-secretion events.

Increased sympathetic activity has effects in systems other than the cardiovascular system. In the liver, glycogen is mobilized and lysed, glucose and potassium are released to the circulation, and the blood sugar rises. Although this effect occurs rapidly, it is a transient one and lasts but thirty to sixty minutes. Beginning at the same time but accelerating slowly to full activity is the hypothalamic-anterior pituitary-adrenocortical axis which greatly increases plasma hydrocortisone from base-line levels of about 15μg to about 50μg-60μg/100 ml. The liver of the anesthetized patient shows a decreased rate of inactivation of hydrocortisone, potentiating the already elevated quantity secreted. The primary action of these higher concentrations of hydrocortisone is to degrade body stores of protein to amino acids, about half of which, by the process of gluconeogenesis, are then converted to glucose within the liver. The net result is an increased hepatic output of glucose which maintains blood sugar levels until the alarm reaction decelerates, usually one or more days postoperatively.

These are normal responses to stress, and they are the responses to general anesthesia. In the normal patient, the rising blood sugar stimulates insulin secretion from the pancreas. The extra insulin limits the extent of the blood sugar rise and disposes of the elevated circulating glucose. The demonstration that ether alters the usual response of blood sugar to insulin indicates that ether has some anti-insulin activity. The anti-insulin activity of other anesthetic agents have not been studied but, in general, the hyperglycemia is not as great with any other currently available anesthetics as with ether.

For the diabetic patient, the ability to secrete insulin is impaired or absent. The stress-induced hyperglycemia is greater and more prolonged than in nondiabetic patients. Though this response is exaggerated, it is not predictable, and the anesthesiologist should perform head-of-the-table tests for blood sugar and acetone levels during anesthesia of the diabetic patient.

The hypoglycemic patient will respond to general anesthesia by developing hypotension. Overinsulinization of the diabetic patient produces not only hypotension under anesthesia but also delays awakening by the patient, or at least such is the interpretation of the persistent hypoglycemic coma.

The patient with thyrotoxicosis or with inadequately suppressed hyperthyroidism will respond to the induction of anesthesia by developing a thyroid crisis. This is an exaggerated response to the epinephrine-thyroxine relationship as it is blocked by pretreatment with reserpine, guanethidine, or high spinal or epidural anesthesia.

Antidiuretic hormone and aldosterone are not secreted in relation to anesthesia per se. However, in the anesthetized patient, blood loss or extracellular fluid translocation results in an exaggerated elaboration of aldosterone and of ADH as physiological efforts to protect the circulating intravascular volume.

REFERENCES

1. Egbert, L.D.; Battit, G.E.; Turndorf, H., and Beecher, H.K.: Value of the preoperative visit by an anesthetist. *JAMA, 185*:553, 1963.
2. Martinez, L.R.; von Euler, C., and Norlander, O.P.: The sedative effect of premedication as measured by catecholamine excretion. *Br J Anaesth, 38*:780, 1966.
3. Giesecke, A.H.; Jenkins, M.T.; Crout, J.R., and Collett, J.M.: Urinary epinephrine and norepinephrine during innovar-nitrous oxide anesthesia in man. *Anesthesiology, 28*:701, 1967.
4. Miller, R.A., and Briscoe, T.J.: Preganglionic sympathetic activity and the effect of anaesthetics. *Br J Anaesth, 37*:804, 1965.
5. Jenkins, G.L.; Livingston, H.L., and Crout,. J.R.: Effects of halothane on catecholamine secretion by the bovine adrenal medulla. *Fed Proc, 28*:355 (Abst. 544), 1969.

Chapter 21

Clinical Aspects of Incapacitating Agents

James S. Ketchum

When Robert Frost wondered in verse whether it would be better to perish by fire or by ice, he could hardly have had in mind the subject of chemical incapacitating agents. Yet the poet laureate's reduction of possible encounters with the infinite to these two varieties of thermal experience captures the essence of the subject of our discussion. For in clinical terms, the broadest generality concerning incapacitating agents is that they almost always work by producing either excitation or depression of the nervous system—a blaze or an icicle, if you will, in the "center" of awareness.

For example, it has been known for centuries—millennia, in fact—that one way to render military opponents temporarily helpless is to induce them to ingest poisonous herbs containing belladonna alkaloids, a class of drugs of which atropine and scopolamine are the best known in today's materia medica. Hannibal allegedly used this trick against unfriendly tribes who threatened his famous march through the Alps 2,000 years ago. In 1672, the Bishop of Muenster's troops employed grenades and projectiles loaded with atropine-like substances during an assault on the city of Groningen. These futuristic devices unfortunately proved to be duds, but this was due to miscalculation, not to a lack of pharmacological creativity. In 1812, a company of hungry French soldiers was rendered *hors de combat* for several days, when they foolishly devoured poisonous berries containing large quantities of belladonna. Some perished from exposure as they wandered about in the bog, half naked and delirious, while others had to be restrained by their comrades as they shouted *aux armes* and attempted to hurl themselves into the campfire. Evidently these well-trained troops, intoxicated by the anticholinergic poison, thought they perceived the enemy moving to the attack amidst the dancing flames. In the twentieth century, the better known incidents involving the bellicose use of belladonna include an unsuccessful attempt by dissident Indo-Chinese officers to poison French troops in 1903, and the placing of atropine in the salt shakers of a cafeteria for employees of Radio Free Europe in the late 1950's. This latter plot also was foiled by prompt discovery.

Today we know that atropine, scopolamine, and their synthetic relatives such as Ditran® cause incapacitation by producing toxic delirium.

Delirium is an old term, but one which has not been improved upon. It refers to a disturbance in brain function manifested by disorientation, confusion, nonsensical repetitious behavior, hallucinations and illusions, and a gross impairment of attention and memory. The picture which comes to mind is that of a staring individual who mumbles to himself, picks at specks on the floor, and fails to comprehend or respond to simple questions or instructions. He is out of touch and unaware. Later he is amnesic for most of his experience. It is a case of incapacitation by ice, a snuffing out of the glow of consciousness, leaving behind an ineffective and ludicrous caricature of human behavior. Delirium produces a man whose behavior suggests he is awake, but who is really asleep; one who is not really on a "trip," but merely undergoing shipment, an automaton who reacts but does not record his reactions, who speaks but does not appreciate his own words, who acts out his dreams for all to see and then forgets them before he is awakened.

But while the production of delirium by depressant drugs such as belladonna enjoys greater historical recognition as a method of causing both large and small scale incapacitation of one's enemies, it is also possible to incapacitate by stimulation, by fire so to speak. Most of us have experienced the jangled nerves and the mild feeling of disorganization which too many cups of coffee can produce when we have unwisely tried to offset lassitude or fatigue by the excessive ingestion of this otherwise refreshing brew. Cocaine, a drug which is no longer used by Western physicians for stimulant purposes, produces a disorganized excitement in those who take large doses. Today, the ability of the amphetamines to produce similar effects is being recognized, and the management of "speed freaks" who are "strung out"—irritable, restless, unable to concentrate, unable to initiate and complete the simple activities required for self-maintenance—has become a formidable problem for many psychiatric treatment centers. And finally, overshadowing these fairly common stimulants, probably because of its more spectacular subjective impact on the psychic apparatus, is the drug LSD.

LSD, like everything else under the sun, is not really new, though at the time of its discovery in 1943 it was thought to be the prototype of a new breed of fantastic chemicals which were variously christened hallucinogenic, psychotomimetic and, most mellifluous of all, "psychedelic" (a term fabricated by Humphrey Osmond in 1956 from Greek words meaning "mind-manifesting"). Subsequently, it was noted that mescaline, the active principle of peyote used in religious ceremonies by American Indians for over a century, produced psychological effects that appeared to be clinically indistinguishable from those produced by LSD. But since it takes 400 mg of mescaline to achieve the incendiary transformation of consciousness that a tenth of a milligram of LSD can spark, one must not be judged too harshly for being somewhat awed by the sheer potency of LSD. After all, the atom bomb is only a bomb, but it still commands a certain respect not accorded the more pedestrian TNT.

The clinical picture of the man whose brain is fired by a superstimulant such as LSD or mescaline is one of mental alertness and physical tension. His emotional state may oscillate between ecstasy as he contemplates the seeming scope and splendor of his heightened awareness and uncanny fear that he will be unable to control the tremendous psychological forces ignited by this chemical spark. The rush of images and ideas is at times overwhelming; it floods over him with an abundance and an effortlessness that is both majestic and terrifying. The mind, working at white heat, conjures up complex designs and geometric landscapes of surrealistic elegance; color and motion are lavishly dispensed by a seemingly inexhaustible generator of kaleidoscopic images; the ear is simultaneously attuned to the rumble of water in pipes, the screech of an automobile down the street, the joking comments of workmen in the courtyard, the crinkle of somebody's cellophane candy wrapper, the muttering of technicians in the hall, and so on—nothing escapes notice, but everything must give way to the next moment of perception. The world vibrates, spins, pulsates, sparkles, and whizzes past, or even stops abruptly. There is no time to do anything but watch and listen—action seems irrelevant or artificial in the face of so rich a mental experience. Memory still functions, but it is difficult to maintain a single purpose or pursue a coherent sequence. In the face of such glorious chaos it is not surprising that irrepressible laughter is sometimes the only adequate commentary, and the enigmatic smile the only suitable reply to a patently absurd question by the uncomprehending doctor. This is incapacitation by pharmacological fire.

In the field exercise conducted some four years ago, a platoon of highly trained troops was declared totally unable to function within an hour after receiving a small dose of LSD. Two of these ordinarily reliable men had climbed trees and would not come down; a third was in a state of panic, praying for deliverance; while still another was laughing uncontrollably and rolling on the ground. The commander of the unit was forced to retire from the field after confessing he had lost control of his men. There is no doubt that mental stimulation can disorganize a military unit as effectively as the production of oblivious stupor.

Five years ago, the most likely candidates for actual incapacitating agents for military use were drugs like LSD on the one hand, and scopolamine or some synthetic relative of greater potency on the other. Today the picture is little changed, with the exception that interest in the anticholinergic type of compound has increased.

A third category of potential agents might include marijuana, which produces effects that are in some ways similar to LSD, though milder. It is a "fire" type of drug, one might say. Yet the fire is a cool one, due to marijuana's relative absence of adrenergic effects. Thus, it is actually a pacifying type of drug which could well remove the will to fight without subjecting the victim to the mental lightning and fireworks which LSD bestows. But efforts to isolate or synthesize the elusive ingredient which emanates from the smoked resin of the marijuana plant have been discouraging. Separate isomers of tetrahydrocannabinol, generally assumed to be the active principle of the natural hashish, have proved relatively ineffective in bringing about the peculiar torpor and extravagant fantasies which characterize the turned-on user of marijuana.

Other classes of drugs are unsuitable for other reasons. The opiates, for example, although they reliably sedate and rob those who take them of physical vigor, are too dangerous because of their low margin of safety. The most potent tranquilizers, such as fluphenazine, can produce lethargy with doses of less than 2 mg, but there is little actual decline in performance capability, and the loss of motivation might conceivably be overcome in an emergency. Straightforward sedatives and hypnotics such as the barbiturates are, in general, insufficiently potent on the one hand and too narrow in their margin of safety to be seriously considered for use as nonlethal weapons on a large scale.

So we are left with two mental conditions which can be considered both incapacitating and attainable within the constraints of safety and feasibility imposed by current policy guidelines. The first, and most likely, is delirium, produced perhaps by a centrally potent anticholinergic agent like scopolamine; and the second is mental overstimulation brought about by a drug of the LSD type.

Differences in the type of effect produced by such chemical agents would, of course, have military implications of a tactical nature. But of greater medical interest are the challenges posed with respect to diagnosis and management.

Diagnosing an isolated case may require differentiation from the so-called functional psychiatric illnesses, but the epidemic character of effects resulting from the large-scale use of incapacitating agents will make it obvious that some sort of an exogenous factor is at work—that is, unless the responsible medical officer himself is a victim of the attack. Assuming he is unaffected, the physician should watch for the characteristic indications of approaching delirium: restlessness, ataxia, drowsiness, and slurred speech are among the early signs. A period of stupor or even coma may supervene, associated with flushing, dry mouth, tachycardia, and temperature elevation. If no specific therapy is administered during this period, a third phase of florid delirium appears, with random primitive movements, stereotyped mumbling, and grasping behavior. Later, hyperactivity with outbursts of disorganized fight-flight-type responses may gradually develop over a period of minutes, hours, or even days, depending on the time course of the specific agent and the route of administration. If percutaneous exposure is involved, for example, it may take as long as forty-eight hours in addition to the ordinary onset time before the development of peak effects. The oral route will delay the onset about an hour. Recovery is generally more gradual than onset, so if a peak of central nervous system depression is seen at eight hours, anywhere from fifty to one hundred hours may be required for complete spontaneous recovery. With shorter acting anticholinergics the total time of action may be as little as two to three hours with peak action at twenty to thirty minutes. Long-acting compounds requiring one to four weeks for recovery are also known.

The inexperienced observer may feel uncertain about his ability to differentiate the intoxication produced by LSD from the delirium resulting from the anticholinergic type of agent. While there are many differences which may be catalogued, diagnosis is to some extent a matter of clinical "feel." The delirious subject impresses one as being generally less aware of things than the examiner, and he is particularly unaware of his own impairment. He may in fact claim he is mentally quite competent in the face of overwhelming evidence to the contrary. If asked to serially subtract 7 from 100, for example, he will often be mumbling something totally unrelated within a few seconds and be unable to recall the original question, but he will resort to ridiculous alibis to distract the examiner and himself from his inability to function. Sooner or later, the examiner gets the feeling that he is talking to something less than a total person, and finds himself adopting a paternalistic or patronizing attitude, or simply not bothering to waste further words with someone who seems unable to retain even simple instructions and explanations.

When confronted with the man who is inebriated with LSD, however, the clinical "feel" is entirely different. The patient may or may not be communicative; often he is not, and answers questions with a few words, or a somewhat irritated or sarcastic comment. He is inattentive only to the extent that he is preoccupied with his own thoughts, but he rarely misses anything that happens or any comment that may pass, for example between the physician and one of his assistants. It is easy to gain his attention, but not necessarily his cooperation. He may appear concerned with issues and questions which transcend the rather banal interests of the examiner. The physician may develop a feeling of being rather stupid and insensitive while trying to conduct an interview with someone who is on LSD. The so-called patient often seems rather amused by the lame efforts the doctor is making to ask meaningful questions. Sometimes unmistakable hostility is displayed and a paranoid suspiciousness that makes the examiner even more self-conscious and uncomfortable. Generally, this type of behavior is not seen in the delirious patient during the early stages of intoxication. Here again, however, individual personality factors play a role in shaping the attitude displayed. Paranoid reactions are, in fact, quite common during the recovery phase of delirium and may occasionally lead to impulsive behavior.

Chemical differentiation between LSD and other drugs is possible though it may not be available as a routine diagnostic test for some years because of the laboratory facilities required. It is, however, quite feasible to measure plasma levels of LSD up to twelve hours following exposure, and requires only a 10-ml sample of plasma which can be frozen for analysis at a later time. The method, developed by Drs. Aghajanian and Bing at these laboratories in 1963, is a fluorometric one, and is based on the little known fact that LSD is among the most fluorescent substances known. With good technique, differences of as little as 1 nanogram can be reliably detected, and since levels of 40 ng or higher are typically attained one to two hours after ingestion of a total dose of 150μg, there is little difficulty in making a positive diagnosis during the first few hours after exposure. Since the biological half-life is about three hours in man, a significant residue should be found even twelve hours following an incapacitating dose. Figure 21–1 illustrates some of these factors and also shows how the LSD level parallels the decrement in performance on an arithmetic task.

Once he has made the diagnosis of chemical incapacitation, what can the physician do to help the situation? For LSD, an accepted treatment for years has been psychological support plus phenothiazine tranquilization, usually with about 50 mg of chlorpromazine. Actually, this is not considered very helpful by many patients, who complain that although they may seem like "Thorazine® on the outside," they are still "LSD on the inside." This, I gather, is a most uncomfortable state of affairs. Nor does Thorazine in doses of 50 or 100 mg have the ability to reverse more than slightly the decrement in cognitive performance produced by LSD, as shown in Figure 21–2. Here we see that subjects receiving LSD followed by placebo treatment were not much more impaired than the experimental groups that received chlorpromazine treatment.

What about delirium produced by the atropine-type drugs? Current textbooks of pharmacology uniformly express the view that the central effects of atropine are not reversible. Fortunately, this is untrue. Over one hundred years ago, an Austrian physician named Kleinwachter published a description of the successful treatment of severe atropine delirium with extract of calabar bean, a potion whose active principle is physostigmine,

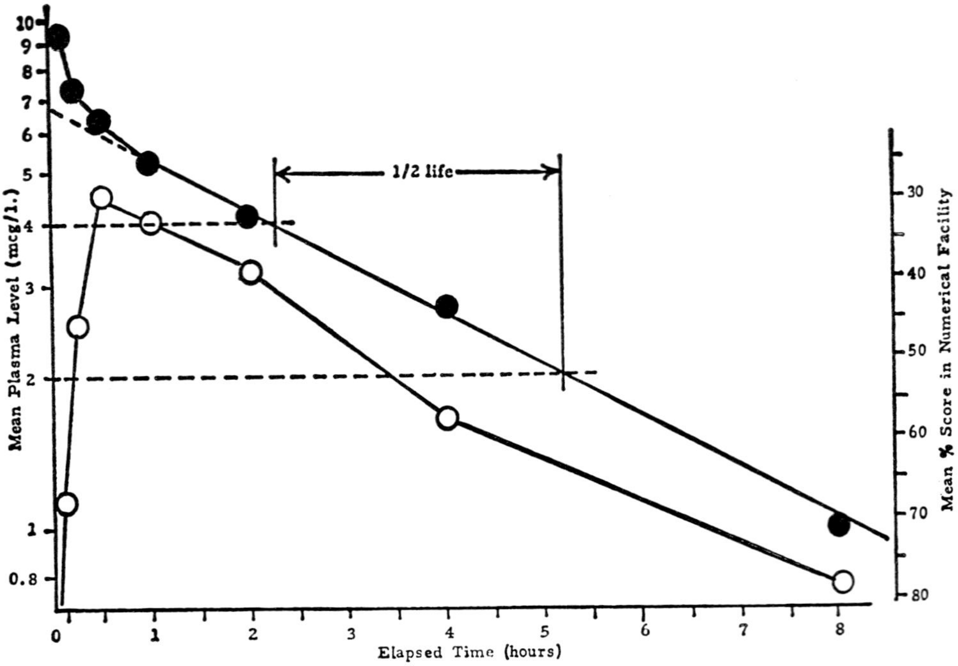

Figure 21–1. Parallelism between plasma concentrations and Number Facility performance scores in human subjects. The solid circles represent mean plasma levels for five subjects who received 2μg of LSD-25/kg of body weight by intravenous injection. The open circles are mean performance scores, expressed as a percentage of predrug baseline. (Reprinted from Aghapanian, G.K., and Bing, O.H.L. Persistence of lysergic acid diethylamide in the plasma of human subjects. *Clin Pharmacol Ther,* 5:611–614, 1964, with the permission of the publisher: The C.V. Mosby Company.)

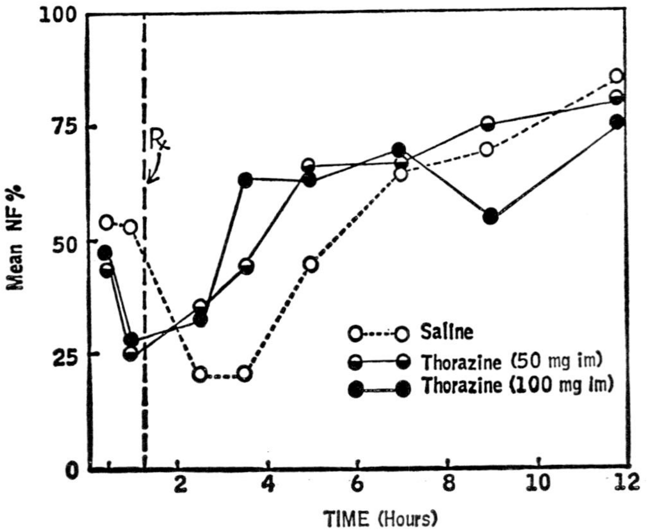

Figure 21–2. Relative ineffectiveness of Thorazine in reversing Number Facility decrement in human subjects. All subjects received 2μg of LSD-25/kg of body weight by intravenous injection. Treatments were administered one hour later. N = 4 for each group.

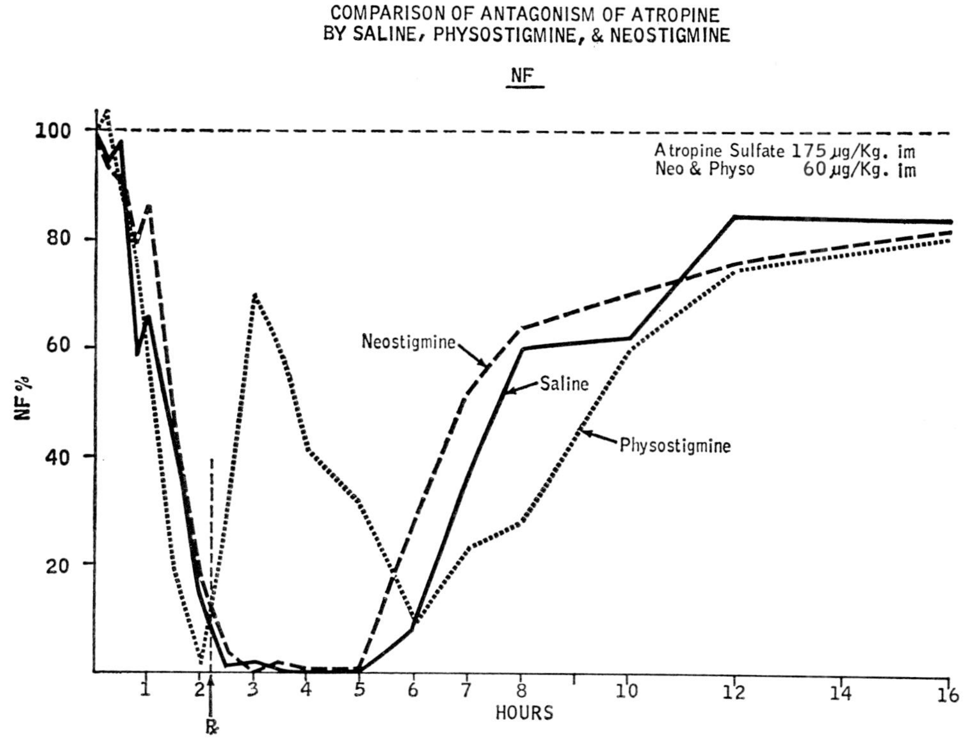

Figure 21–3. Atropine: Reversal of Number Facility decrement by physostigmine but not by neogstigmine. All subjects received 175μg of atropine sulfate/kg of body weight by intramuscular injection. N = 4 for each group.

a familiar short-acting cholinesterase inhibitor. Figure 21–3 shows that, more than a century following Kleinwachter's unheeded recommendation that further tests be done with this promising drug, it has not lost its therapeutic effectiveness. A single dose of 4 mg of physostigmine, given near the time of peak atropine effects on cognitive performance, produces nearly complete reversal of the cognitive decrement within twenty minutes, but in the absence of repeated doses, the scores again decline after an hour or two. Figure 21–4 shows similar results with scopolamine, again with a single dose of physostigmine. It is interesting to note that if the physostigmine is given prior to the time of peak scopolamine action, it appears to have less therapeutic effect. No explanation has been found for this relative refractoriness. The responses of a single subject who received Ditran followed by physostigmine (Fig. 21–5) indicate that this synthetic cholinergic blocking agent also responds similarly to treatment with appropriate doses of physostigmine, despite claims to the contrary by some investigators.

Physostigmine treatment requires close supervision, and to be optimal, an hourly dosage of 2 to 5 mg (not more than 4 mg in a single dose) may

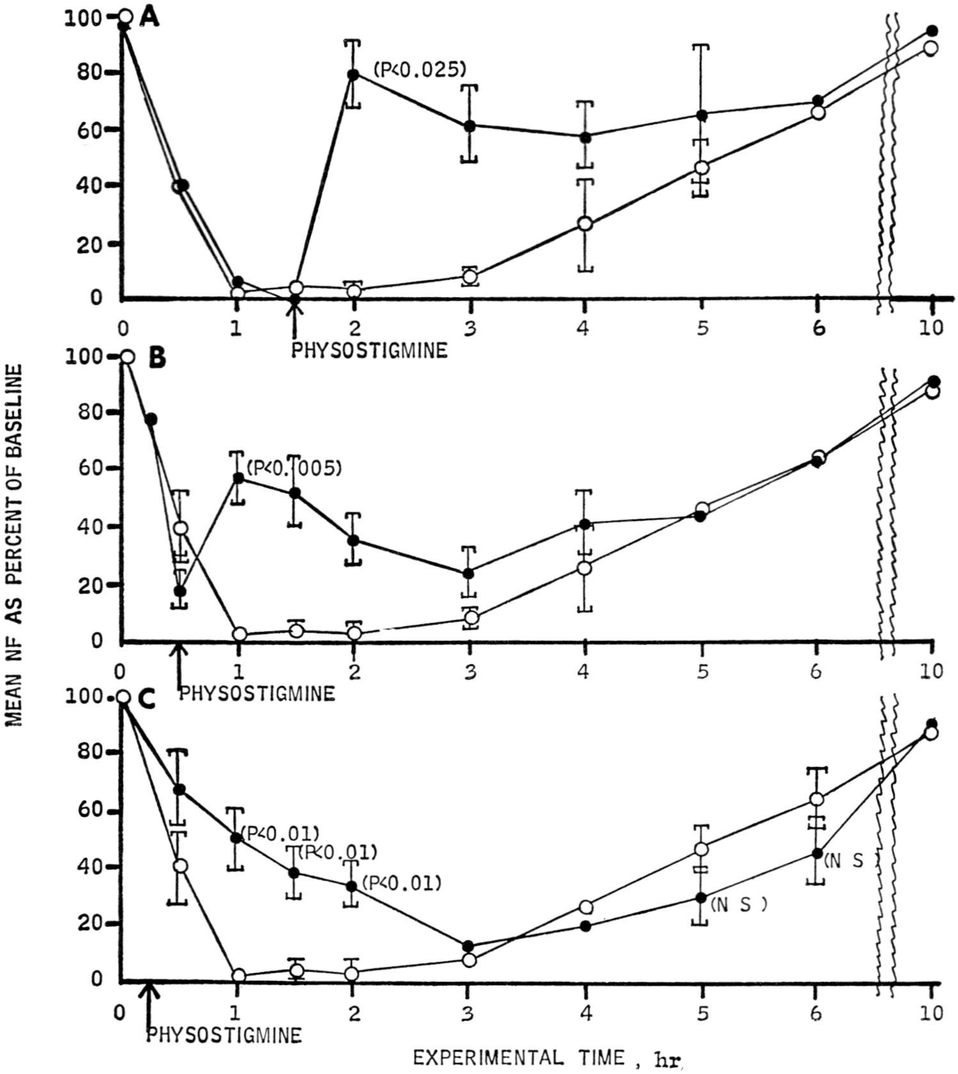

Figure 21–4. Scopolamine. Reversal of Number Facility decrement by physostigmine. (Reprinted from: Crowell, E.B., Jr., and Ketchum, J.S. The treatment of scopolamine-induced delirium with physostigmine. *Clin Pharmacol Ther*, 8:409–414 (1967) with permission of the publisher: The C.V. Mosby Company.

be required. As in the too zealous treatment of diabetic coma, where sugar must sometimes be given to counteract excessive insulin, excess physostigmine may require titration with small doses of atropine. This, however, is seldom necessary if the patient is monitored closely.

In summary, incapacitating agents with which we may be confronted probably will prove to be either central excitants such as LSD or delirients such as atropine, scopolamine, or synthetic anticholinergic drugs. Historical

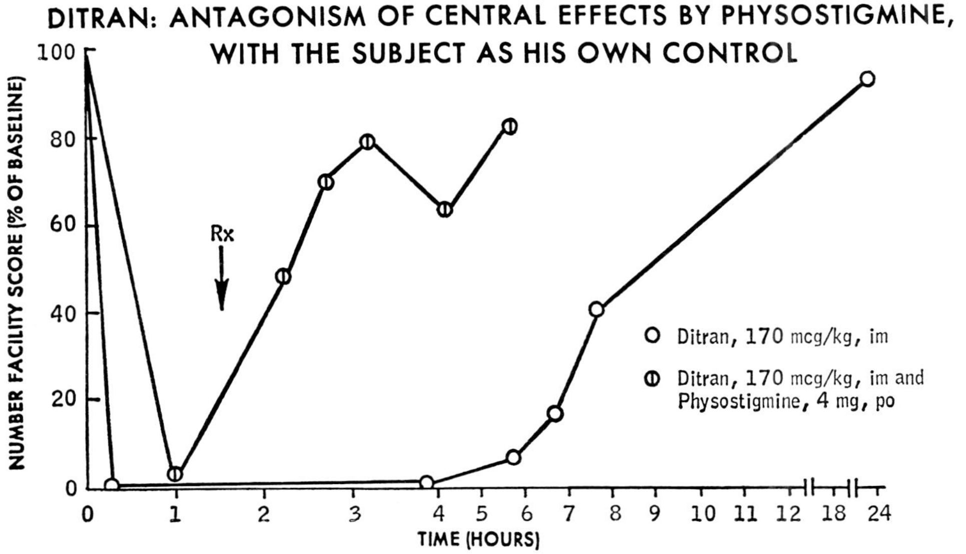

Figure 21–5. Ditran: Reversal of Number Facility decrement by physostigmine.

precedent for such weapons is quite extensive and the belladonna alkaloids have always been favored for such purposes. Differential diagnosis depends on the ability to recognize the oft-neglected syndrome of delirium with its hallucinations, incoherence, and confusion. This condition is quite different from the hyperalert perplexity characteristic of LSD. A blood test for LSD may establish the diagnosis in doubtful cases. Treatment for LSD intoxication is supportive. No specific antagonist has yet been demonstrated. For anticholinergic delirium specific therapy can be carried out by "titration" with physostigmine. If incapacitating agents are ever used on a mass scale, it may be necessary for the physician to place greater emphasis on physical restraint than on specific pharmacological therapy. Diagnosis may be easier, but the complexities of the management problem can hardly be predicted in detail.

In a war fought with incapacitating agents, we must be prepared for fire as well as ice. For incapacitation lies at either extreme of the human condition, and the physician should learn how to diagnose and treat hypothermia and hyperpyrexia of the psyche as confidently as he does the physiological chills and fever of more familiar illnesses.

Chapter 22

Comment on Plasma Insulin and HGH

TSUTOMU OYAMA

In normal man under non-stressful conditions, it is known that the administration of insulin induces hypoglycemia which in turn increases plasma HGH (anterior pituitary growth hormone) level. On the other hand, hyperglycemia evokes the decrease of plasma HGH levels. In the diabetic patient the magnitude of elevated plasma growth hormone caused by hypoglycemia is lower than that in normal man.

It is also well known that excessive concentrations of growth hormone can produce impaired glucose tolerance and insulin resistance. It has been postulated that the lipolytic effect of growth hormone, by which an alternate fuel is made available, may be a part of body defenses to carbohydrate deprivation. Growth hormone can also be regarded as a stress hormone.

For these reasons, we studied the effects of ether anesthesia, methoxyflurane anesthesia and the effect of operation on blood levels of insulin, HGH, glucose and free fatty acid (FFA) in 40 patients (20 for ether and 20 for methoxyflurane). The measurements of blood levels of insulin and HGH were made by radioimmunoassay. In these cases normal saline was infused and whole blood was given during the procedure when needed. Anesthesia alone continued for 45 minutes before starting operation.

During the first 45 minutes of ether anesthesia, there was no significant change in plasma insulin level. However, at the end of 1 hour of operation, there was a significant decrease in plasma insulin level as compared with pre-induction level. The elevation in blood sugar during ether anesthesia was significant, and it significantly rose further during the operation.

The blood HGH levels slightly but significantly increased during ether anesthesia alone, and it was markedly elevated during the operation. The blood FFA levels were slightly increased during the procedure.

We demonstrated the significant increase of HGH level at the 30th or 45th min after the starting of penthrane anesthesia alone, and a peak was made at the 30 min after starting the operation. But blood insulin level showed insignificant variations during anesthesia.

There is a Japanese proverb saying that the more a child sleeps, the more he grows up! Takahashi, et al. demonstrated that a plasma HGH peak appeared with the onset of deep sleep at night and it decreased to almost zero during daytime. But peak HGH secretion is delayed or impaired if the onset of sleep is delayed or interrupted.

The increasing effect of anesthesia on blood HGH level is far less than that with natural sleep. Furthermore the increasing effect of methoxyflurane on the blood HGH was more prominent than in ether. These points are important and are worthy of further study.

REFERENCES

1. Oyama, T., and Takazawa, T.: Effects of diethyl ether anaesthesia and surgery on carbohydrate and fat metabolism in man. *Canad Anaesth Soc J, 18*:51–59, 1971.
2. Oyama, T., and Takazawa, T.: Effect of methoxyflurane anaesthesia and surgery on human growth hormone and insulin levels in plasma. *Canad Anaesth Soc J, 17*:347–358, 1970.
3. Takahashi, Y.; Kipnis, D.M., and Daughaday, W.H.: Growth hormone secretion during sleep. *J Clin Invest, 47*:2079–2090, 1968.

Chapter 23

Comment on Thyroxine

Tsutomu Oyama

Effects of anesthesia and surgery on serum thyroxine levels were measured in 148 surgical patients. Exogenous blood levels were determined by using radioactive thyroxine (I^{131}—T_4) in 41 patients and endogenous blood thyroxine levels were measured by using radioactive tri-iodothyronine (I^{131}—T_3) sponse uptake test in 107 patients.

Ether anesthesia increased exogenous blood thyroxine levels markedly. However, halothane anesthesia increased blood thyroxine level only slightly; thiopental and methoxyflurane anesthesia did not raise blood thyroxine levels at all.

It is suggested that the increased blood thyroxine level under ether anesthesia is due to dominant alteration in the distribution of liver thyroxine. Thiopental failed to elevate either blood thyroxine concentration or thyroxine metabolism.

This lack of action of thiopental is attributable to the chemical structure of thioureylene nucleus—that is the thiobarbiturate nucleus, thioureylene NH—C(S)—NH, is chemically allied to the anti-thyroid agent, thiouracil, and pentobarbital has Oxygen instead of Sulfur, and this accounts for the non-antithyroid property. This was demonstrated in an animal experiment by Oyama.

The effects of anesthesia on endogenous blood thyroxine level was similar to that of exogenous thyroxine. After 40 minutes of anesthesia with ether and halothane before the operation started there were significantly increased endogenous serum thyroxine concentrations, but thiopental, N_2O and O_2 anesthesia definitely decreased the thyroxine level. Methoxyflurane anesthesia did not significantly increase blood thyroxine levels. Surgical operation further elevated the serum thyroxine level in both ether and halothane anesthesia, but no significant rise of thyroxine levels was observed in the thiopental—N_2O—O_2 or methoxyflurane groups. The effect of spinal and epidural anesthesia do not increase blood thyroxine level during anesthesia prior to operation nor during the surgery. The results of our study suggest that thiopental followed by N_2O and O_2 or methoxyflurane would be a better choice than ether or halothane for the patient with hyperthyroidism.

The cardiac irritability in the hyperthyroid patient is elevated, and this is due to the overactivity of myocardial beta-receptor by the increased thyroxine, which indicates that beta-blockade is effective for arrhythmias.

REFERENCES

1. Oyama, T.; Shibata, S., and Matsuki, A.: Thyroxine distribution during ether and thiopental anesthesia in man. *Anesth & Analg, 48:*1–6, 1969.
2. Oyama, T.; Shibata, S.; Matsuki, A., and Kudo, S.: Serum endogenous thyroxine levels in man during anaesthesia and surgery. *Brit J Anaesth, 41:*103–108, 1969.
3. Oyama, T.; Shibata, S.; Matsuki, A., and Kudo, T.: Effect of methoxyflurane anaesthesia on thyroid function in man. *Canad Anaes Soc J, 16:*204–208, 1969.
4. Oyama, T., and Matsuki, A.: Serum levels of thyroxine in man during spinal anesthesia and surgery. *Anesth & Analg, 50:*309–313, 1971.
5. Oyama, T., and Matsuki, A.: Serum levels of thyroxine in man during epidural anesthesia and surgery. *Der Anaesthesist 19:*298–300, 1970.

Author Index

Subject Index

D

E

N

Q

R